INTERNATIONAL CENTRE FOR MECHANICAL SCIENCES

COURSES AND LECTURES - No. 53

KURT MAGNUS
TECHNICAL UNIVERSITY OF MUNICH

GYRODYNAMICS

COURSE HELD AT THE DEPARTMENT
OF GENERAL MECHANICS
OCTOBER 1970

UDINE 1974

SPRINGER-VERLAG WIEN GMBH

Originally published by Springer-Verlag Wien-New York in 1972

ISBN 978-3-211-81229-7 ISBN 978-3-7091-2878-7 (eBook)

DOI 10.1007/978-3-7091-2878-7

PREFACE

The peculiar motions of spinning bodies have always fascinated mathematicians, physicists and engineers. Indeed, the literature relating to the problems of spinning bodies has long since grown to such an extent as to become almost unsurveyable. It may therefore be useful and desirable to summarize thy most important results on the theory of spinning bodies and, what is more, to present them from some unified points of view. In doing so it is primarily intended to create a kind of bridge between the classical theory of spinning bodies and the more recent results obtained in the course of applications in technology, navigation and space travel. Although individual gyroscopic intruments will not be discussed in the course of these lectures, it is nevertheless proposed to examine the most important general phenomena that occur in them, including the effects of oscillation, rectifying effects, and problems of tuning.

In preparing the texts of these lectures I have made use of some sections of my book "Kreisel, Theorie und Anwendungen" (Gyros, Theory and Applications) published in 1971 by the publishing house of J. Springer, Berlin. I am therefore indebted to the publishers for the consideration they have shown me in making possible the preparation and publication of these greatly abbreviated lecture notes.

I should also like to express my thanks to the International Centre of Mechanical Sciences and, more particularly, to its untiring and ever-fertile director, Professor Dr. Luigi Sobrero.
My thanks are due not only for the suggestion to hold this course of lectures at the Centre in Udine about my special field of work, but also for the unfailing help and support that has been given in realizing this project.

K. Magnus

Udine, October 1970

Magnus

1. Introductive Remarks.

In these notes the term "gyro" is used to describe, quite generally, a rigid body that is performing rotatory (or spinning) motions. The term is therefore intended to cover both the "rigid bodies with a fixed point" of the classical theory of spinning bodies and the fast-spinning gyro wheels of gyroscope technology. However, unlike the frequent practice in the case of technical applications, the term "gyro" is not intended to describe either a rotor enclosed in a housing or, even less so, a complete instrument that contains a fast-spinning rotor.

The modes of motion of a gyro are not capable of being deduced from an analogy with the translational motions of point masses, this being due to the fact that the gyro is characterized by a peculiar anisotropy vis-à-vis its rotatory movements. While the inertia of a point mass can be described, quite simply, by means of a scalar magnitude, its mass, in the case of a rigid body it becomes necessary to use a tensor for this purpose. The types of motion of a gyro depend on the shape of the body, on the forces and torques that act on it, on the constraints that the environment imposes on it, on the type and the motion of the reference system, and lastly on the initial state of motion. Variations of these magnitudes will lead to a multiplicity of forms of motion that it is almost impossible to survey, and only a few of these, either because they are typical or because they are important for applications, can be treated here.

Following Duschek and Hochrainer [1] , the analytical notation will be used for the purpose of mathematical description. Vectors will therefore be characterized by the apposition of an index (e.g. X_i, F_j,

ω_k), tensors of the second degree by means of two indices (e.g. a_{ij}, c_{jk}, θ_{lm}). In general these indices can have the values 1, 2, 3, but the available store of values could be greater in the case of matrices. Important symbols used in the next text are the unit vector e_i (for which $|e_i| = 1$), the "Kronecker symbol" δ_{ij}, and the "Levi-Civita symbol" ϵ_{ijk}. The scalar product of the vectors x_i and y_i will be represented by $z = x_i y_i$, and their vectorial product by means of $z_i = \epsilon_{ijk}\, x_j y_k$. $y_i = b_{ij} x_j$ describes a linear vector function. Where a suffix occurs twice, the summation must always be effected over the whole range of available values. In particular, we have $x_i = \delta_{ij} x_j$.

The relative position of two coordinate systems 1, 2, 3 and 1′, 2′, 3′ can be described by means of the matrix $a_{ij} = \cos \alpha_{ij}$ of the direction cosine. In this connection, for example, α_{12} represents the angle through which the 1-axis of the original system must be rotated in order to be brought into the position of the 2 -axis of the other system. The following relationships apply in the case of this transformation

(1) $$a_{ij}\, a_{ik} = a_{ji}\, a_{ki} = \delta_{jk}$$

(2) $$x_i = a_{ij}\, x'_j \;;\; x'_i = a_{ji}\, x_j$$

(3) $$T_{ij} = a_{ik}\, a_{jl}\, T'_{kl} \;;\; T_{ij}' = a_{ki}\, a_{lj}\, T_{kl}$$

(4) $$\epsilon_{ijk}\, \epsilon_{ilm} = \begin{vmatrix} \delta_{jl} & \delta_{jm} \\ \delta_{kl} & \delta_{km} \end{vmatrix} = \delta_{jl}\, \delta_{km} - \delta_{kl}\, \delta_{jm} .$$

2. Fundamentals.

2.1. Geometry of Masses.

The tensor of inertia of a rigid body can be represented in the form

$$\theta_{ij} = \int (x_k x_k \delta_{ij} - x_i x_j)\, dm = \begin{bmatrix} A & -F & -E \\ -F & B & -D \\ -E & -D & C \end{bmatrix} \tag{5}$$

Both the moments of inertia A, B, C and the products of inertia D, E, F here appear as the constitutent elements of the tensor of inertia. They are subject to the frequently used inequalities

$$A + B \geqslant C \;;\; B + C \geqslant A \;;\; C + A \geqslant B \,, \tag{6}$$

$$A \geqslant 2D \;;\; B \geqslant 2E \;;\; C \geqslant 2F \,. \tag{7}$$

Although both the moments of inertia and the products of inertia depend on the choice of the coordinate system to which reference is made, the tensor θ_{ij} itself is a magnitude that is independent of the reference system and characterizes the physical inertia properties of the body. For this reason, the transformation formulae (3) can be applied to θ_{ij} whenever it is desidered to change from one reference system to another. This can be demonstrated as follows. Using the transformation (2), i.e. $x_i' = a_{ji} x_j$, it follows from (5) that

$$\theta_{ij}' = \int (x_k' x_k' \delta_{ij}' - x_i' x_j')\, dm =$$

$$= \int (a_{lk} a_{mk} x_l x_m \delta_{ij}' - a_{ki} a_{lj} x_k x_l)\, dm \,. \tag{8}$$

Since

$$a_{lk}\, a_{mk} = \delta_{lm} \;;\; \delta_{lm}\, x_l = x_m \;;\; a_{ki}\, a_{lj}\, \delta_{kl} = \delta'_{ij}$$

(8) can be transformed into

$$\theta'_{ij} = a_{ki}\, a_{lj} \int (x_m x_m \delta_{kl} - x_k x_l)\, dm = a_{ki}\, a_{lj}\, \theta_{kl} \;. \tag{9}$$

This corresponds to the transformation law (3) ; the tensor property of θ_{ij} is therefore proved.

When the origin of the reference system is changed from O to P (Fig. 1), we have $x_i = r_i + y_i$; substituting this in (5), we then obtain

$$\begin{aligned} \theta_{ij} &= \int (y_k y_k \delta_{ij} - y_i y_j)\, dm + m\,(r_k r_k \delta_{ij} - r_i r_j) + \\ &+ 2\delta_{ij}\, r_k \int y_k\, dm - r_i \int y_j\, dm - r_j \int y_i\, dm \;. \end{aligned} \tag{10}$$

The last three terms of this expression disappear when the reference point P coincides with the centre of gravity S of the body. In this case we obtain

$$\theta^{O}_{ij} = \theta^{S}_{ij} + m\,(r_k r_k \delta_{ij} - r_i r_j) \,. \tag{11}$$

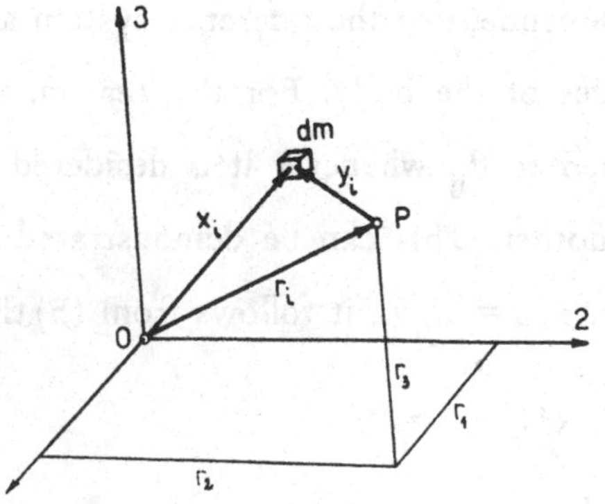

Fig. 1

This is the general form of the so-called Huygens-Steiner theorem. This theorem states that the moments of inertia for any given reference system with the origin at O consist of the corresponding values for a parallel system of axes with the origin at the centre of gravity S and those components that come into being when the whole of the mass m of the body is assumed as being concentrated in S.

The moment of inertia θ for any given axis that has been rotated with respect to the reference system can be obtained as follows : we take the axis under consideration as the 1′-axis of a rotated reference system and then obtain from (9)

$$\theta = \theta'_{11} = a_{k1} a_{l1} \theta_{kl} \; . \tag{12}$$

This is a quadratic form of the direction cosine of the 1′-axis. Using $a_{i1} = (a_1, a_2, a_3)$, as well as equation (5), (12), is then transformed into

$$\theta = Aa_1^2 + Ba_2^2 + Ca_3^2 - 2Da_2 a_3 - 2Ea_3 a_1 - 2Fa_1 a_2 \; . \tag{13}$$

If we now introduce the vector $y_i = \rho a_{i1}$ into the above, where

$$\rho = \frac{1}{k} = \sqrt{\frac{m}{\theta}} \tag{14}$$

is the module of inertia of the body for the direction under consideration, then (12) becomes transformed into

$$\theta_{ij} \, y_i y_j = m \tag{15}$$

or, writing this in the extended form,

$$Ay_1^2 + By_2^2 + Cy_3^2 - 2Dy_2 y_3 - 2Ey_3 y_1 - 2Fy_1 y_2 = m \; . \tag{16}$$

Since we have a constant magnitude on the right-hand side of the above expression, the quadratic form (16) defines an ellipsoid along whose surface the terminal point of the vector y_i will move when the direction of the 1′-axis is varied. The principal axes of this ellipsoid, the so-called Poinsot ellipsoid, are the principal axes of inertia. If these are chosen as the reference axes, the products of inertia D, E, F will disappear. Using the principal moments A′, B′, C′, the equation of the ellipsoid then becomes transformed into

$$A'y_1^2 + B'y_2^2 + C'y_3^2 = m . \tag{17}$$

The principal moments of inertia can also be chosen as eigenvalues of the eigenvalue equation

$$(\theta_{ij} - \lambda\delta_{ij})\, a_j = 0 \tag{18}$$

The principal axes can then be obtained as those values of the unit vector a_j for which there exist solutions of equation (18).

2.2. Classification of Gyros.

On the basis of the relationships that exist between the principal moments of inertia of a body, that is to say, according to the shape of the ellipsoid of inertia of the body, the various gyros can be classified as follows :

1) A body with A = B = C is called a spherical gyro. Its ellipsoid of inertia is a sphere, but its external shape does not by any means have to be spherical. Homogeneous cubes or tetrahedrons, for example, are spherical gyros.
2) If any two of A, B, C are of equal magnitude, the ellipsoid of inertia of the body is rotationally symmetrical. The body is then described as a symmetrical gyro. All homogeneous rotating bodies are symmetrical gyros with respect to reference points that lie on the axis of symmetry. The axis of symmetry is also referred to as the axis of the figure.
3) When A, B, C are all different from each other, one speaks of an unsymmetrical gyro. The appropriate ellipsoid of inertia has three axes.

In the case of a spherical gyro all the axes passing through the reference point are equally entitled to be regarded as principal axes. In the case of a symmetrical gyro, on the other hand, all the axes lying in the equatorial plane, i.e. at right angles to the axis of symmetry or the axis of the figure, have equal rights.

Among the symmetrical gyros one can further distinguish between

2a) elongated gyros - in this case we have $A = B > C$ (rod in the direction of the 3-axis) – and

2b) flattened gyros, where we may have, for example, $A = B < C$ (symmetrical plate in the 1, 2-plane).

Similar descriptions have also been used in the case of the unsymmetrical gyro. If $A > B > C$, then the gyro is

3a) short-axled with respect to the 1-axis (this corresponds to the flattened gyro),

3b) medium-axled with respect to the 2-axis, and

3c) long-axled with respect to the 3-axis (this corresponds to the elongated gyro).

An overall view of the various types of gyros could be obtained, for example, by using the following three representations :

1) Plotting the principal moments of inertia A, B, C as the sides of a plane triangle, as shown in Fig. 2. If A is assumed to be constant, then the possible point representing the vertex of the triangle formed by the sides A, B, C can lie anywhere within the shaded half-plane.

2) Plotting in the B/A, C/A – plane. In view of the inequalities (6), the points representing the various possible types of gyros can lie anywhere within the shaded half-strip shown in Fig. 3.

3) A representation in the shape-triangle, this latter being obtained as follows:

A, B, C are plotted as lengths along the axes of a Cartesian coordinate system (Fig.4). Each possible ellipsoid of inertia will then be characterized by a point P lying in the first octant. Moreover, the points representing bodies with similar ellipsoids of inertia will lie on the line joining P to the origin O. Since we are not interested in the absolute magnitude of the ellipsoid of inertia but only in its shape (or form), it will be sufficient to consider the points of intersection D of the various lines OP with the plane defined by

$$A + B + C = 1$$

Given the governing inequalities (6), the possible points D representing the various forms of the ellipsoid of inertia must then lie within the shaded triangle of Fig. 4. This triangle is known as the shape-triangle.

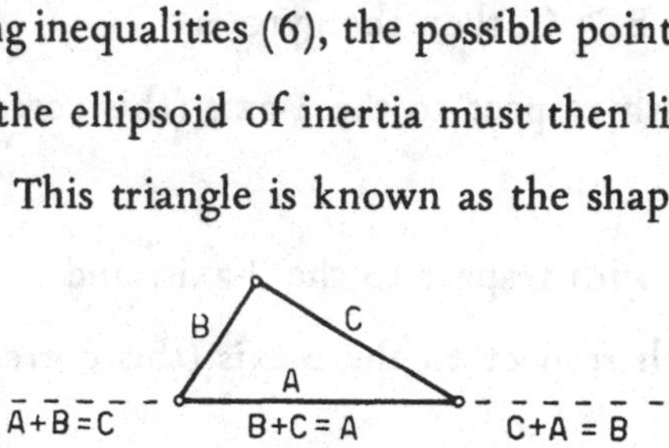

Fig. 2

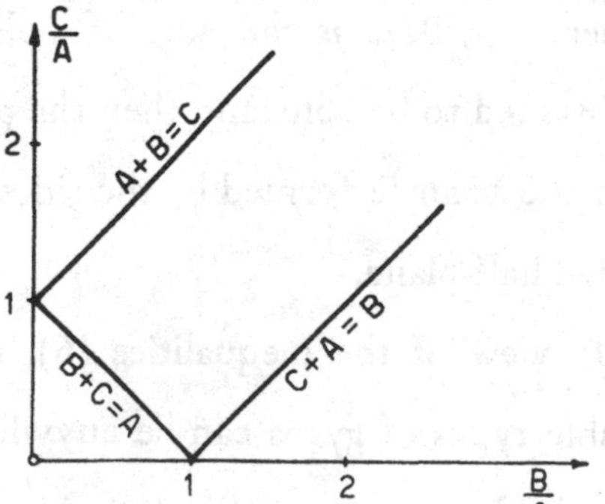

Fig. 3

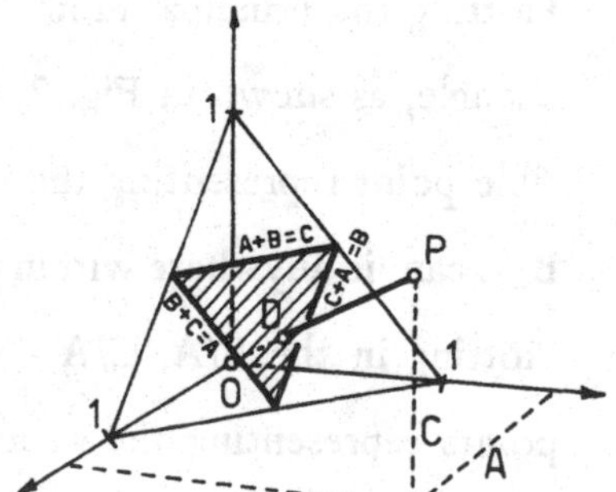

Fig. 4

These three types of representations are shown side by side in Fig. 5. In this figure corresponding points are always characterized by means of the same number and, consequently, corresponding limiting lines are always designated by means of the same set of three numbers. The various points are to be interpreted as follows :

Point 1 : A = 0, rod in the 1-direction

Point 2 : B = 0, rod in the 2-direction

Point 3 : C = 0, rod in the 3-direction

Point 4 : B = C = A/2, symmetrical plate in the 2, 3-plane

Point 5 : C = A = B/2, symmetrical plate in the 3, 1-plane

Point 6 : A = B = C/2, symmetrical plate in the 1, 2-plane

Point 7 : A = B = C, spherical gyro.

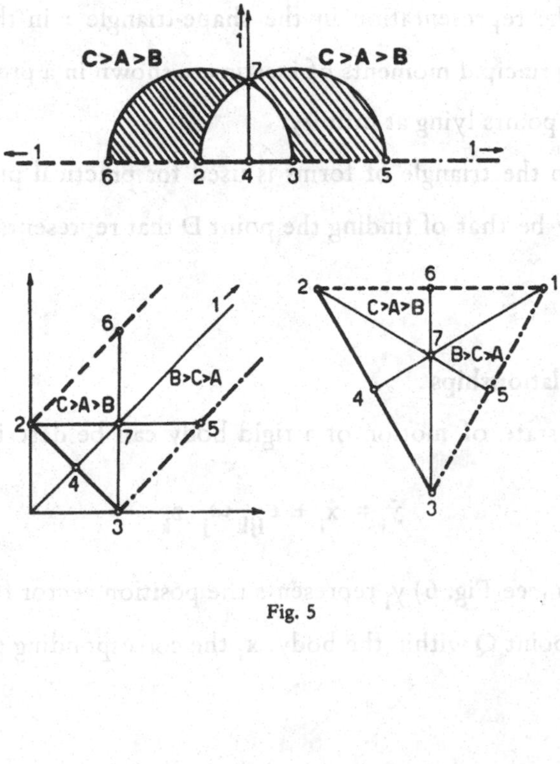

Fig. 5

Points lying on lines 1-7 (or, respectively, on lines 2-7 or 3-7) represent elongated gyros with respect to the 1-axis (or, respectively, the 2- or 3-axis), while those lying on lines 7-4 (or, respectively, lines 7-5 or 7-6) represent symmetrical gyros with respect to the 1-axis (or, respectively, the 2- or 3-axis).

Points lying on the straights 2-4-3 (or, respectively, 3-5-1 or 2-6-1) represent plates in the 2, 3-plane (or, respectively, the 3, 1- or 1, 2-planes).

The limiting lines 2-7-5, 3-7-6, 1-7-4 subdivide the possible domain of the representative points into six partial domains, each of which corresponds to a particular sequence of magnitudes for A, B, C. The appropriate magnitude sequence for two of these domains have been shown. The domains corresponding to medium-axled gyros with respect to the 1-axis have been shaded.

A comparison of these three types of representations shows that they are topologically equivalent. In the case of complicated problems, however, one should opt for the representation in the shape-triangle : in this representation none of the three principal moments of inertia are shown in a preferential manner, nor are there any points lying at infinity.

When the triangle of forms is used for practical purposes, the problem will generally be that of finding the point D that represents a concrete set of values of A, B, C.

2.3 Kinematic Relationships.

The state of motion of a rigid body can be described by means of

$$\dot{y}_i = \dot{x}_i + \epsilon_{ijk}\, \omega_j\, z_k \tag{19}$$

In this expression (see Fig. 6) y_i represents the position vector from a fixed point O to any chosen point Q within the body, x_i the corresponding position vector to a reference point P that is fixed with respect to the body, and ω_i the vector of the angular velocity of the body. The vectors x_i and ω_i can be interpreted as the vector components of a kinemate with which the body is being moved. When $\dot{x}_i \parallel \omega_i$, the kinemate becomes reduced to a screw of motion.

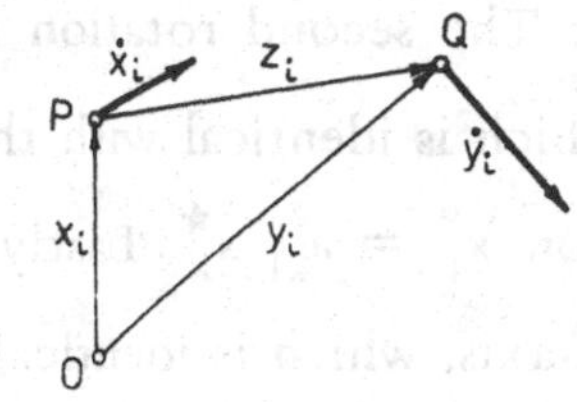

Fig. 6

A geometric description of the motion of a rigid body with a fixed point can be given by considering a polar cone fixed with respect to the body and rolling down a trace cone, this latter being fixed with respect to space. The momentary axis of rotation will then always lie on the line of contact between the two cones.

For the purposes of the analytical description of the motions of rigid bodies, one can make use of the direction cosines of the various angles of the axes or of generalized angular coordinates, for example, the Eulerian angles ψ, ϑ, θ or the cardanic angles α, β, γ. In both cases it is possible to describe the rotation of the 1′2′3′– system (this system being fixed with respect to the body) relative to the 1 2 3 – system (this latter system being fixed with respect to space) by means of three successive rotations about axes that are kept fixed in each case. When Eulerian angles are being used, this procedure can be schematically indicated by

$$(1\,2\,3) = \underset{3\,\equiv\,3^*}{\textcircled{\psi}} \Rightarrow (1^*\,2^*\,3^*) = \underset{1^*\equiv\,1^\circ}{\textcircled{\vartheta}} \Rightarrow (1^\circ\,2^\circ\,3^\circ) = \underset{3^\circ\equiv\,3'}{\textcircled{\varphi}} \Rightarrow (1'\,2'\,3') \qquad (19a)$$

The first rotation is carried out in the positive direction around the 3-axis, an axis that is identical with the 3^*-axis of an intermediate system. This rotation is described by the transformation $x_k^* = a_{jk}^{\psi}\, x_j$, where the matrix a_{jk}^{ψ} depends only on

ψ. The second rotation is carried out in a positive direction around the 1-axis, which is identical with the 1°-axis. This rotation is described by the transformation $x_l^\circ = a_{kl}^{\vartheta} x_k^\star$. Lastly, the system is rotated through the angle φ around the 3°-axis, which is identical with the 3′-axis and therefore fixed with respect to the body itself. This is described by $x_i' = a_{li}^{\varphi} x_i^\circ$. Considering these three successive rotation as a whole, we obtain

$$(20) \qquad x_i' = a_{li}^{\varphi} x_l^\circ = a_{kl}^{\theta} a_{li}^{\varphi} x_k^\star = a_{jk}^{\psi} a_{kl}^{\theta} a_{li}^{\varphi} x_j = a_{ji} x_j$$

where

$$(21) \quad a_{ji} = \begin{bmatrix} \cos\psi\cos\varphi - \sin\psi\cos\theta\sin\varphi & -\cos\psi\sin\varphi - \sin\psi\cos\theta\cos\varphi & \sin\psi\sin\theta \\ \sin\psi\cos\varphi + \cos\psi\cos\theta\sin\varphi & -\sin\psi\sin\varphi + \cos\psi\cos\theta\cos\varphi & -\cos\psi\sin\theta \\ \sin\theta\sin\varphi & \sin\theta\cos\varphi & \cos\theta \end{bmatrix}$$

In a completely corresponding manner, a description with the help of cardanic angles can be schematically represented by

$$(1\,2\,3) = \textcircled{\alpha} \Rightarrow (1^\star 2^\star 3^\star) = \textcircled{\beta} \Rightarrow (1^\circ 2^\circ 3^\circ) = \textcircled{\gamma} \Rightarrow (1'\,2'\,3') \qquad (20a)$$
$$1 \equiv 1^\star \qquad\qquad 2^\star \equiv 2^\circ \qquad\qquad 3^\circ \equiv 3'$$

The overall transformation is then given by

$$(22) \qquad x_i' = a_{li}^{\gamma} x_l^\circ = a_{kl}^{\beta} a_{li}^{\gamma} x_k^\star = a_{jk}^{\alpha} a_{kl}^{\beta} a_{li}^{\gamma} x_j = a_{ji} x_j$$

where

$$(23) \quad a_{ji} = \begin{bmatrix} \cos\beta\cos\gamma & -\cos\beta\sin\gamma & \sin\beta \\ \cos\alpha\sin\gamma + \sin\alpha\sin\beta\cos\gamma & \cos\alpha\cos\gamma - \sin\alpha\sin\beta\sin\gamma & -\sin\alpha\cos\beta \\ \sin\alpha\sin\gamma - \cos\alpha\sin\beta\cos\gamma & \sin\alpha\cos\gamma + \cos\alpha\sin\beta\sin\gamma & \cos\alpha\cos\beta \end{bmatrix}$$

In the case of gyro problems it is often useful to employ reference systems that are not fixed with respect to space, but rather systems that are either fixed with respect to the body or generally in motion. In the absence of translatory displacements, the relationship between the coordinates of a vector x_i in the fixed system and a vector x_k' in the moving system is given by the transformation equation

$$x_i = a_{ik} x_k' \quad .$$

From this we obtain the change of the vector as

$$\frac{dx_i}{dt} = \frac{d}{dt}(a_{ik} x_k') = a_{ik}\frac{dx_k'}{dt} + \frac{da_{ik}}{dt} x_k' \quad . \tag{23a}$$

The first term of this expression represents the relative change of x_k' with respect to the moving system. This term is frequently abbreviated as

$$a_{ik}\frac{dx_i'}{dt} = \frac{d'x_i'}{dt} \quad . \tag{23b}$$

With the help of the angular velocity vector ω_j, which represents the speed of rotation of the moving system with respect to the fixed one, the second term can be expressed as follows

$$\frac{da_{ik}}{dt} x_k' = \epsilon_{ijk}\omega_j x_k' \quad . \tag{23c}$$

This expression represents the carrying that is impressed upon the vector by virtue of the fact that it is being carried away by the moving system. The absolute change is therefore equal to the sum of the relative change and the carrying change :

(24) $$\frac{dx_i}{dt} = \frac{d'x_i'}{dt} + \epsilon_{ijk}\,\omega_j x_k' \,.$$

2.4. Kinetic Relationships

The kinetic energy of a rigid body

$$T = \frac{1}{2}\int \dot{y}_i\,\dot{y}_i\,dm$$

can be brought into the form

(25) $$2T = m\dot{x}_i\,\dot{x}_i + 2m\dot{x}_i\,\epsilon_{ijk}\,\omega_j z_k^s + \int \epsilon_{ijk}\,\omega_j\, z_k\,\epsilon_{ilm}\,\omega_l\, z_m\,dm \,.$$

by making use of equation (19). The last term describes the kinetic energy of the rotation T_D. By means of (4) this term can be transformed into

(26) $$\begin{aligned} 2T_D &= \int(\delta_{jl}\delta_{km} - \delta_{kl}\delta_{jm})\,\omega_j\omega_l z_k z_m\,dm = \\ &= \int(\omega_j\,\omega_j\, z_k z_k - \omega_j\omega_k z_j\, z_k)\,dm = \\ &= \omega_i\omega_j \int(z_k z_k\,\delta_{ij} - z_i\, z_j)\,dm = \theta_{ij}\,\omega_i\,\omega_j \,. \end{aligned}$$

Just like the kinetic energy, the angular momentum of a rigid body, given by

$$H_i^o = \int \epsilon_{ijk}\, y_j\, \dot{y}_k\, dm$$

can likewise be subdivided into three parts with the help of $y_i = x_i + z_i$ and equation (19). We thus obtain

$$H_i^\circ = m\epsilon_{ijk} x_j \dot{y}_k^s + m\epsilon_{ijk} z_j^s \dot{x}_k + \int \epsilon_{ijk} z_j \epsilon_{klm} \omega_l z_m dm \quad . \tag{27}$$

In this expression the upper index S indicates magnitudes that refer to the centre of mass. With the help of equation (4) the component of the angular momentum caused by the rotatory motion can be transformed into

$$\begin{aligned} H_{D_i}^\circ &= \int \epsilon_{ijk} z_l \epsilon_{klm} \omega_l z_m dm = \\ &= \int (\delta_{il}\delta_{jm} - \delta_{jl}\delta_{im}) \omega_l z_j z_m dm = \\ &= \int (z_j z_j \delta_{il} \omega_l - z_l z_i \omega_l) dm = \theta_{il}^P \omega_l . \end{aligned} \tag{28}$$

In this expression θ_{il}^P is the tensor of inertia for the point P which is fixed with respect to the body (see Fig. 4). Using (26), it follows from (28) that

$$T_D = \frac{1}{2} \theta_{ij} \omega_i \omega_j = \frac{1}{2} H_{Di} \omega_i \quad . \tag{29}$$

The most important of the fundamental laws of the theory of spinning bodies, the theorem of angular momentum, can be written in the form

$$\frac{dH_i^\circ}{dt} = M_i^\circ \tag{30}$$

for a reference system (fixed with respect to space) with the fixed point O. In the case of a moving reference system with the reference point P fixed with respect

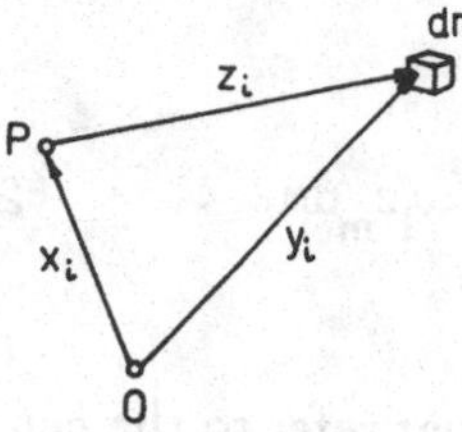

Fig. 7

to the body (Fig. 7), the theorem of angular momentum can be transformed into

$$(31) \qquad \frac{d}{dt}\left(\theta^P_{ij}\,\omega_j\right) + m\epsilon_{ijk}\, z^s_j\, \ddot{x}_k = M^P_i \; .$$

It can readily be seen that there are two cases when, in spite of the moving reference point P, the above expression reduces to the form (10), i.e. :

1) when $\ddot{x}_k = 0$, that is to say, when P is not being accelerated, and
2) when $z^s_j = 0$, that is to say, when $P \equiv S$ and is therefore also the centre of mass.

For the purpose of solving gyro problems, it is often advantageous to make use of a reference system that is fixed with respect to the body and whose origin coincides with the fixed point. In this case, with the help of equation (24), the theorem of angular momentum (30) can be brought into the form

$$(32) \qquad \frac{d'H_i}{dt} + \epsilon_{ijk}\,\omega_j H_k = M_i$$

Since in $H_i = \theta_{ij}\,\omega_j$ the tensor of inertia for a system fixed with respect to the body is constant, equation (32) can also be written as follows :

$$(33) \qquad \theta_{ij}\frac{d'\omega_j}{dt} + \epsilon_{ijk}\,\omega_j\,\theta_{kl}\,\omega_l = M_i \; .$$

The relationships (32) and (33) are vector forms of the classical Eulerian equations for spinning bodies. In this form they are also valid for a single rigid body with a fixed point. When calculating gyro systems, however, these equations have to be further generalized because the parts of the system do not generally have any fixed points.

Numerous problems of spinning bodies can also be solved with the help of the equations of motion developed by Lagrange. In the case of conservation systems with a potential energy U, these equations assume the well known form

$$\frac{d}{dt}\left(\frac{\partial T}{\partial \dot{x}_\nu}\right) - \frac{\partial T}{\partial x_\nu} + \frac{\partial U}{\partial x_\nu} = 0, \quad (\nu = 1 \ldots n) \tag{34}$$

In this expression all the n generalized coordinates must be successively used for x_ν.

3. The Interactions of Forces and Motions in a Gyro.

In the case of gyros that are subject to the action of any kind of forces, we are normally interested in two types of problems : we either have to calculate the forces that are acting when certain types of motions are observed, or we have to determine the motions that occur as the result of the action of known forces. Problems of the first kind can generally be solved without difficulty, because the torques appear in the equations of motion in a linear form. On the other hand, up to the present time it has only been possible to find exact solutions for very few cases of the second type. This is due to the fact that the components of the angular velocity occur in the equations of motion in quadratic form and also as derivatives with respect to time.

The effect of moving gyros can be calculated directly from the theorem of angular momentum (30). The torque M_i to be introduced into this expression is exerted upon the gyro from outside. When $M_i = 0$, then the vector of the angular momentum H_i remains constant in both magnitude and direction. On the other hand, if the gyro is externally driven in such a manner that H_i changes, then the gyro will exert a gyro torque

$$(35) \qquad M_i^K = - \frac{dH_i}{dt}$$

upon the bearings. For the practical calculation of this torque it will be best to transform to a rotating system. If ω_j is the vector of the angular velocity of this system, then it follows from (32) that

$$(36) \qquad M_i^K = - \frac{dH_i}{dt} = - \frac{d'H_i}{dt} - \epsilon_{ijk}\,\omega_j\,H_k \quad .$$

If the amount of H_i remains unchanged, then ω_j can always be chosen in such a way that H_i remains constant in the moving system, that is to say $d'H_i/dt = 0$. In this case we have quite simply

$$(37) \qquad M_i^K = \epsilon_{ijk}\,H_j\,\omega_k \quad .$$

Formulae (36) and (37) are used for the calculation of the gyro effects of machinery rotors and can also serve for the purpose of interpreting the behaviour of the so-called curving top.

3.1. Gyros influenced by Gravity.

This detonation is applied to gyros when the external torque acting on the body is due to the force of gravity. If the position of the centre of gravity S of the body with respect to the fixed point F is given by the vector s_j (see Fig. 8), then the torque is given by

$$M_i = \epsilon_{ijk} s_j G_k = G \epsilon_{ijk} a_{3j} s_k \qquad (38)$$

where the unit vector a_{3j} is oriented in the direction of the 3-axis, which is fixed with respect to space and assumed to be vertical, and $G_k = -Ga_{3k}$ is the vector of the force of gravity. The equation of motion of the gravity-influenced gyro then follows from equation (32) as

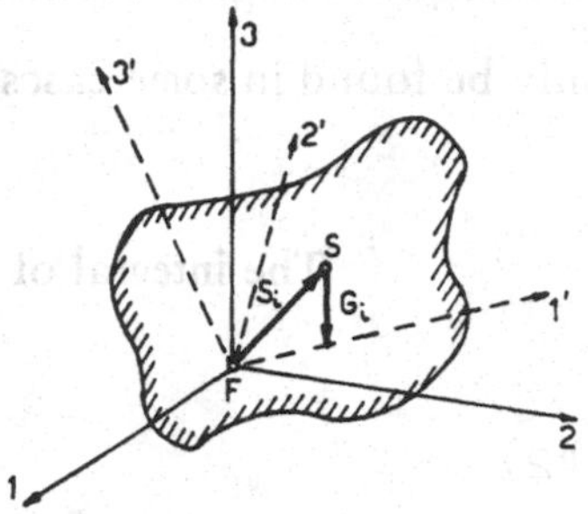

Fig. 8

$$\frac{d'H_i}{dt} + \epsilon_{ijk} \omega_j H_k = G \epsilon_{ijk} a_{3j} s_k \ . \qquad (39)$$

The unknowns in this equation are the vector ω_j, which characterizes the state of motion, and the vector a_{3i}, which determines the position of the system with respect to the vertical (the system, of course, being fixed with respect to the body). But equation (39) is not yet sufficient for calculating the motion of the gyro. For this purpose we must also make use of the kinematic relationship

$$\frac{da_{3i}}{dt} = \frac{d'a_{3i}}{dt} + \epsilon_{ijk} \omega_j a_{3k} = 0 \qquad (40)$$

This equation expresses the fact that a_{3i} is a vector that remains fixed with respect to space. Since a_{3i} is a unit vector, its coordinates are such that

$$a_{31}^2 + a_{32}^2 + a_{33}^2 = 1 \ . \qquad (41)$$

No completely general solution of the equations of motion of gyros influenced by gravity has so far been found. Such a solution would be possible if one could succeed in finding three first integrals in the sense of the Jacobian theory of integration. In general, however, it is only possible to state two integrals, an integral of angular momentum and an integral of energy. A third integral could only be found in some cases.

The integral of the angular momentum can be obtained from

$$\frac{dH_i}{dt} = G\,\epsilon_{ijk}\,a_{3j}s_k \tag{42}$$

after performing a scalar multiplication with a_{3i}. The right-hand side of this equation disappears, since $\epsilon_{ijk}a_{3j}s_k a_{3i} = 0$. Moreover, since $da_{3i}/dt = 0$, the left-hand side of the expression can be transformed into

$$\frac{dH_i}{dt}\,a_{3i} = \frac{d}{dt}(H_i\,a_{3i}) = 0\,. \tag{42a}$$

or

$$H_i\,a_{3i} = H_0 = \text{const.} \tag{43}$$

In other words, the vertical component of the angular momentum remains constant.

The energy integral can be obtained from (39) after performing a scalar multiplication with ω_i. Since

$$\frac{d'H_i}{dt}\,\omega_i = \frac{d'}{dt}\left(\frac{1}{2}\,\theta_{ij}\,\omega_i\,\omega_j\right) = \frac{d'T}{dt}\;, \tag{43a}$$

and

$$\epsilon_{ijk} a_{3j} s_k \omega_i = \epsilon_{ijk} \omega_j a_{3k} s_i = - \frac{d' a_{3i}}{dt} s_i = - \frac{d'}{dt} (a_{3i} s_i)$$

one finds, after integration with respect to time, that

$$T + U = \frac{1}{2} \theta_{ij} \omega_i \omega_j + G s_i a_{3i} = E_0 = \text{const.} \qquad (44)$$

In other words, the sum of the kinetic energy and the potential energy remains constant.

The table on the next page shows all the cases for which it has so far been possible to find exact solutions of the equations of motion of the gravity-influenced gyro. Without any exception, these are all special cases in which the form of the ellipsoid of inertia, or the position of the centre of gravity, or the initial conditions, are subject to certain limitations. Indeed it will be seen that in some cases the imposed limitations apply to two (and sometimes even all three) of these.

Case No.1 has been included only for the sake of completeness : it concerns the case of the gyro on which no forces are acting. General solutions of the gyro equations, i.e. solutions that are valid for any initial conditions, are known only for Cases 2 and 3. Cases 4 to 9 are concerned with motions that can only come into being when there are particular initial conditions, and at times some very special ones. The fact that the Cases 1, 2 and 3 listed in the table occupy a special position is also demonstrated by a theorem that has been proved by Ljapunov : these three cases are the only ones where the components of the vectors ω_i and a_{3i} become unique functions of time no matter what the initial conditions may be. Particular interest also attaches to Case 7, because these motions, which were discovered by Staude, remain valid for any gyro whatsoever, no mat-

	Limitations for the			
	shape of the ellipsoid of inertia	position of the centre of gravity	initial conditions	defected by
1	arbitrary	(small s)=0	arbitrary	Euler Poinsot
2	A = B	$s_1 = s_2 = 0$ $s_3 \neq 0$	arbitrary	Lagrange Poisson
3	A=B=2C	$s_1 \neq 0$ $s_2 = s_3 = 0$	arbitrary	Kovalew-skaja
4	A=B=4C	$s_1 \neq 0$ $s_2 = s_3 = 0$	$(H_i a_{3i})_o = 0$	Gorjačev Čaplygin
5	A=B=4C	$s_1 \neq 0$ $s_2 = s_3 = 0$	$(\epsilon_{ijk} \omega_j a_{k3})_o = 0$	Mercalov
6	2A = C	$s_1 = s_2 = 0$ $s_3 \neq 0$	$(\omega_i a_{i2})_o = 0$	Steklov
7	arbitrary	arbitrary	$(\epsilon_{ijk} \omega_j a_{3k})_o = 0$	Staude
8	arbitrary	$\frac{s_1}{s_3} = \sqrt{\frac{C(A-B)}{A(B-C)}}$ $s_2 = 0$	$(H_i s_i)_o = 0$	H e s s
9	arbitrary	$\frac{s_1}{s_3} = \sqrt{\frac{A-B}{B-C}}$ $s_2 = 0$	$\omega_o = \sqrt[4]{\frac{4 s^2}{(A-B)(B-C)+(A-B+C)^2}}$	Grioli

ter what its shape or the position of its centre of gravity may be.

The literature of the classical theory of spinning bodies is almost exclusively concerned with gravity-influenced gyros. A great deal of effort has been dedicated to discovering cases for which the non-linear equations of motion (39) could be solved in an exact manner. However fascinating the results of this effort may be for a mathematician, one must nevertheless point out that they are only of rather scarce interest from a physical point of view, and even less so from the point of view of gyroscope technology. One should add that, ever since we have had at our disposal efficient electronic computers, the reduction of an equation to a series of solvable integrals is no longer of such vital importance as, quite rightly, attached to it within the framework of classical mechanics. Nowadays we experience no difficulty in calculating the motions of a gravity-influenced gyro by means of numerical integration, and this to any desired degree of accuracy and for any given initial conditions. As regards the special results that have been obtained for the classical problem of the gyro influenced by gravity, the reader should refer to the rather extensive literature on this specialized subject [for example 2, 3, 4].

3.2. Self-excited Gyros.

In accordance with Grammel's definition, a gyro is described as self-excited if its motions are produced or maintained by torques M_i whose components are known in the reference system fixed with respect to the spinning body. These M_i may be constant, but they may also appear as functions of time or of the angular velocity. However, they must not be dependent in any way on the body's momentary position in space. In these conditions it becomes possible to calculate the state of motion of the gyro solely on the basis of the Eulerian kinetic equations, i.e. one does not have to make the parallel effort of integrating the kinematic equations, a process that generally involves a great deal of toil.

The torque vectors M_i can be functions of time as regards both their magnitude and their direction. For example, both these are of interest in the positional adjustement of spaceships, where the correcting torques are provided by means of steering jets with an adjustable thrust. The loss of mass of the body caused by jet propulsion is normally neglected when considering such problems. This is undoubtedly admissible in the case of attitude adjustments, but it would become problematical when examining the wobbling motions of rising rockets.

3.2.1. Self-excited symmetrical Gyros.

When A = B and the excitation torques are functions of time, the equations of motion of the self-excited gyro are as follows :

$$(45)\quad \left.\begin{aligned} A\dot{\omega}_1 - (A-C)\,\omega_2\omega_3 &= M_1\ (t) \\ A\dot{\omega}_2 + (A-C)\,\omega_3\omega_1 &= M_2\ (t) \\ C\dot{\omega}_3 \qquad\qquad\qquad &= M_3\ (t) \end{aligned}\right| \begin{aligned} &, \\ &, \\ &. \end{aligned}$$

It follows from the third of these equations that

$$(46),\qquad \omega_3 = \omega_{30} + \frac{1}{C}\int M_3\ (t)\ dt.$$

Introducing a new independent variable α by means of the transformation

$$(47)\qquad d\alpha = \omega_3\ dt$$

the first two equations of (45) can now be brought into a linear form that readily lends itself to integration. If we now characterize the derivatives with respect to the variable α by means of a vertical dash, i.e. (′), we can write

$$\dot{\omega} = \frac{d\omega}{dt} = \frac{d\omega}{d\alpha}\,\frac{d\alpha}{dt} = \omega'\omega_3 \; . \tag{47a}$$

It thus follows from (45) that

$$\left.\begin{aligned} \omega_1' - a\omega_2 &= \frac{M_1}{A\omega_3} = m_1 \;, \\ \omega_2' + a\omega_1 &= \frac{M_2}{A\omega_3} = m_2 \;, \end{aligned}\right| \tag{48}$$

The abbreviation a = (A–C)/A has been used in these equations.

In view of the inequalities (6), the value of a will always lie within the range $-1 \leqslant a \leqslant +1$. The case of the spherical gyro with a = 0 can be excluded from these considerations, because in this case the system (45) can be solved in an elementary manner.

If we now introduce the complex magnitudes

$$\omega_1 + i\omega_2 = \omega^\star \;; \quad m_1 + im_2 = m^\star \tag{48a}$$

then the equations (48) can be reduced to

$$\omega^{\star\prime} + ia\omega^\star = m^\star \tag{49}$$

but, in view of the transformation (47), the complex excitation $m^\star = m^\star(t)$ must now be represented as a function of the new variable α.

The general solution of (49) with the constant of integration $\omega_o^\star = \omega^\star(O)$ is given by

$$\omega^\star = e^{-ia\alpha}\left[\omega_o^\star + \int_o^\alpha m^\star(\beta)\, e^{ia\beta}\, d\beta\right] \tag{50}$$

This, together with (46), yields the state of motion for any given self-excitation function.

3.2.2. Excitation by means of Torque Pulses.

The case of excitation by means of torque pulses of short duration is of interest for the purposes of practical applications in space travel. If we suppose that the torques have no components in the direction of the axis of symmetry, then, if $M_3 = 0$, it immediately follows that $\omega_3 = \omega_{3o} =$ constant.

A sequence of torque pulses of short duration can be represented with the help of the Dirac function $\delta(t)$ in the form

$$(51) \qquad M(t) = \sum_\nu M_\nu \, \delta(t-t_\nu)$$

In this expression t_ν is the moment of time at which the νth pulse occurs, and M_ν is a measure of the magnitude of this pulse. Corresponding expressions are obtained for the expressions m(t) according to (48), so that the general solution (50) can now be written in the form

$$(52) \quad \omega^\star = \omega_o^\star e^{-ia\omega_{30} t} + \omega_{30} \int_0^t e^{-ia\omega_{30}(t-\tau)} \sum_\nu m_\nu^\star \delta(\tau - t_\nu) \, d\tau$$

Each individual pulse changes the angular velocity by an additional amount of

$$(52a) \qquad \Delta\omega_\nu^\star = \omega_{30} \, e^{-ia\omega_{30}(t-t_\nu)} \, m_\nu^\star \int_{t_\nu-\epsilon}^{t_\nu+\epsilon} \delta(\tau - t_\nu) \, d\tau \, .$$

With the help of the unit step function

$$(52b) \qquad 1(t-t_\nu) = \int_0^t \delta(\tau - t_\nu) \, d\tau = \begin{cases} 0 & \text{for } t < t_\nu \\ 1 & \text{for } t > t_\nu \end{cases}$$

the general solution (52) can be brought into the form

$$\omega^{\star} = \omega_0^{\star}\, e^{-ia\omega_{30} t} + \omega_{30} \sum_{\nu} m_{\nu}^{\star}\, 1(t-t_{\nu})\, e^{-ia\omega_{30}(t-t_{\nu})} . \qquad (53)$$

This can be interpreted as a superposition of eigenoscillations (nutations) that are caused by the pulses. Between any two pulses the vector ω_i travels round a part of a polar cone, and this cone, since $M = 0$, is a straight circular cone that has the axis of symmetry of the gyro as its axis. Each pulse can change the aperture of the polar cone. The polar curve can thus be calculated step by step and this, in turn, makes it possible to calculate the trajectory.

3.2.3. Unsymmetrical Gyros.

No general solutions have so far been found for self-excited unsymmetrical gyros. All the same it has been possible to solve a number of special problems, this being principally due to Grammel [6]. By way of example, we shall here examine the case of self-excitation by means of a constant torque in one of the principal axes.

When $M_1 = M_{1o} \neq 0$ and $M_2 = M_3 = 0$, the Eulerian equations of motion become transformed into

$$\left.\begin{aligned} A\dot{\omega}_1 - (B - C)\, \omega_2 \omega_3 &= M_{10} , \\ B\dot{\omega}_2 + (A - C)\, \omega_3 \omega_1 &= 0 , \\ C\dot{\omega}_3 - (A - B)\, \omega_1 \omega_2 &= 0 . \end{aligned}\right| \qquad (54)$$

If one assumes $A > B > C$, then the differences that occur will be positive. Using the transformation $d\alpha = \omega_1 dt$, $\omega = \omega' \omega_1$, the last two of these equations become transformed into

$$(55)\qquad \begin{aligned} B\omega_2' + (A - C)\omega_3 &= 0 \ , \\ C\omega_3' - (A - B)\omega_2 &= 0 \ . \end{aligned}$$

If the initial conditions are $\omega_2 = \omega_{2o}$, $\omega_3 = \omega_{3o}$, the general solution of these equations is given by

$$(56)\qquad \begin{aligned} \omega_2 &= \omega_{20}\cos\nu\alpha - \sqrt{\frac{(A-C)C}{(A-B)B}}\,\omega_{30}\sin\nu\alpha \ , \\ \omega_3 &= \omega_{30}\cos\nu\alpha + \sqrt{\frac{(A-B)B}{(A-C)C}}\,\omega_{20}\sin\nu\alpha \ , \\ \nu &= \sqrt{\frac{(A-B)\ (A-C)}{BC}} \ . \end{aligned}$$

If one substitutes these results in the first of equations (54), one obtains

$$(56a)\qquad \dot{\omega}_1 = \frac{M_{10}}{A} + \frac{(B-C)}{A}\,\omega_2(\alpha)\ \omega_3(\alpha) = F(\alpha) \ .$$

Multiplying the above by $\omega_1 = \dfrac{d\alpha}{dt}$ yields

$$(56b)\qquad \dot{\omega}_1\,\omega_1 = \frac{d}{dt}\left(\frac{\omega_1^2}{2}\right) = F(\alpha)\frac{d\alpha}{dt}$$

and this can now be integrated to obtain

$$(57)\qquad \omega_1 = \frac{d\alpha}{dt} = \sqrt{\omega_{10}^2 + 2\int F(\alpha)\,d\alpha} \ .$$

A further integration will then yield

$$t = t_0 + \int \frac{d\alpha}{\sqrt{\omega_{10}^2 + 2\int F(\alpha)\, d\alpha}} = t(\alpha) \tag{58}$$

The required solution $\omega_i(t)$ can now be obtained by introducing the reverse function $\alpha = \alpha(t)$ in (57) and (56). For a discussion of the solution the reader is referred to Grammel [6].

3.3. Externally excited Gyros

The term external excitation is used when the external torques appearing in the gyro equations are known as functions of time in the reference system fixed with respect to space. Periodic functions are of primary interest in this case, but stochastic functions also become interesting for the purposes of technical applications. External excitations of this type occur in atomic and molecular physics when rotating polarized particles become subject to the influence of alternating fields. External excitations in mechanics generally occur when a gyro is shaken by a tremor. If the point of suspension of a gyro becomes accelerated by $b_i(t)$, a force of $-m[b_i(t) - g_i]$ will act at the centre of mass, and this force depends on the overall acceleration. If the centre of mass S is separated from the point of suspension O by a distance s_i, and if the direction OS is chosen as 3′-axis fixed with respect to the body, then we can write $s_i = sa_{i3}$. If the reference system fixed with respect to space possesses a vertical 3-axis, we also have $g_i = -ga_{3i}$. The external torque acting on the gyro is then given by

$$M_i = -ms\epsilon_{ijk} a_{j3} [b_k(t) + ga_{3k}] . \tag{59}$$

According to the nature of the accelerating function $b_k(t)$ and the direction of the tremor we can here have a multiplicity of cases, but only a few of these have so far been analyzed. Two of these cases will now be mentioned.

When the tremor acts on the point of suspension in a vertical direction, and when

$$b_k(t) = a_{3k} b_o \cos\Omega t \tag{60}$$

that is to say, when

$$b_1 = b_2 = 0 \; ; \; b_3(t) = b_0 \cos\Omega t$$

then it follows from equations (59) that

$$M_i = -msg\,(1 + \frac{b_o}{g} \cos\Omega t)\,\epsilon_{ijk}\, a_{j3}\, a_{3k} \tag{61}$$

This torque lies in the horizontal plane and is proportional to $\sin\vartheta$, when ϑ is the angle between the vertical (a_{3k}) and the 3′-axis fixed with respect to the body (a_{j3}). If the 3′-axis is also the axis of symmetry of the gyro (that is to say, when A = B), then the equations of motion can be transformed into a non-linear differential equation of the form

$$\dot{\vartheta} = F\,(\vartheta, t) \tag{62}$$

in which periodic coefficients occur. Weidenhammer [7] succeeded in solving this equation by means of an expansion into series and in so doing discovered two novel types of effects :

1) The aperture of the precession cone changes with respect to the classical case of a gyro with a fixed point of suspension. This is analogous to the shift of the equilibrium position of a pendulum when its point of suspen

sion becomes subject to a tremor.

2) Changed conditions of stability occur in this case, and g has thus to be replaced by the expression $g\,(1 + \zeta \cos \vartheta_o)$. In this expression $\zeta = msb_o^2/2gA\Omega^2$ represents a dimensionless parameter of the tremor.

When the point of suspension of a symmetrical gyro becomes subject to tremors and there are small deviations of the axis of symmetry from the vertical direction, the equations of motion can be be brought into the form

$$\left.\begin{aligned} A\ddot{\alpha} + C\omega_{30}\dot{\beta} - ms\,(g+b_3)\,\alpha &= msb_2 \quad , \\ A\ddot{\beta} - C\omega_{30}\dot{\alpha} - ms\,(g+b_3)\,\beta &= -msb_1 \quad . \end{aligned}\right| \qquad (63)$$

In these equations α represents the angle of rotation of the gyro round the 1-axis, and β the corresponding angle round the 2-axis, both axes being fixed with respect to space. If we now write

$$x = \alpha + i\beta \qquad b^{\star} = b_1 + ib_2$$

both the equations (63) can be combined to the complex form

$$A\ddot{x} - iC\omega_{30}\,\dot{x} - ms\,(g+b_3)\,x = -imsb^{\star} \,. \qquad (64)$$

When $b_3 = O$, the horizontal tremor will be exclusively determined by the excitation function $b^{\star}(t)$. If we put $b^{\star} = b_o e^{i\Omega t}$, we obtain an excitation that rotates with a frequency Ω, while an excitation in the direction of one of the axes can be described, for example, by means of

$$b^{\star} = b_0 \cos\Omega t = \frac{1}{2}\, b_0\,(e^{i\Omega t} + e^{-i\Omega t}) \qquad (64a)$$

When the above is used as the excitation function, the general solution of (63) can be brought into the form

(65) $$x = K_1 e^{i\omega_N t} + K_2 e^{i\omega_P t} + R_1 e^{i\Omega t} + R_2 e^{-i\Omega t} .$$

where ω_N and ω_P represent the eigenfrequencies of the gyro (nutation and precession), while R_1 and R_2 are resonance functions. When projected into the complex plane, the solution curve based on (65) is found to consist of the superposition of four circular motions. According to the various frequency conditions one can obtain some very different types of trajectories for a point on the axis of symmetry of the gyro. Fig. 9 illustrates the changes in the trajectories of the resonance components that occur when the excitation frequency Ω is varied. In this connection we have $s > o$ ($s < o$) for a gyro whose centre of mass lies above (below) the point of suspension.

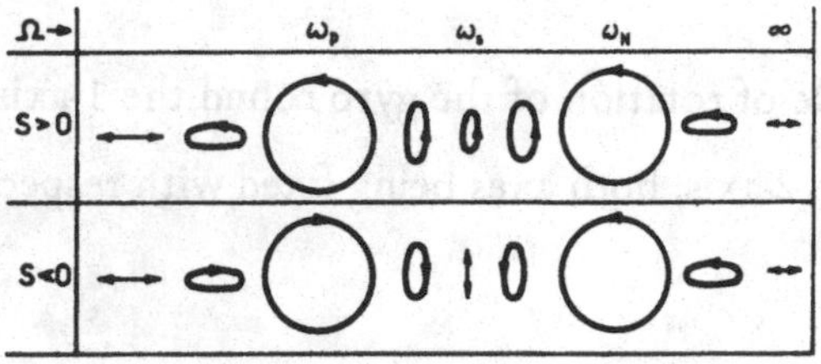

Fig. 9

More general solutions of equation (64), for example for elliptically polarized excitations or for non-harmonic excitation functions, can be obtained in a well-known manner.

4. Gyrostats and Cardanic Suspended Gyros.

4.1. Gyrostats.

Following Lord Kelvin, the term gyrostat is used to describe a rigid body that contains in its interior a symmetrical rotor. This latter can rotate about an axis that is fixed with respect to the body and it therefore has one degree of freedom vis-à-vis the enveloping body. In this case it is assumed that the rotation of the rotor relative to the body takes place about the axis of symmetry of the body itself and that, consequently, the mass distribution of the overall system, i.e. of the gyrostat as a whole, is not in any way changed as a result of the rotations of the rotor. It is therefore possible to dèfine constant moments of inertia for the gyrostat system consisting of two bodies and then to calculate the motions of the gyrostat, the procedure being rather similar to that used in the case of a single rigid body.

4.1.1. The Equations of Motion of the Gyrostat.

If H_i^K represents the vector of the angular momentum of the enveloping body (box or housing) and H_i^R the vector of the angular momentum of the rotor, then the theorem of angular momentum can be applied to the overall momentum $H_i = H_i^K + H_i^R$ and we thus obtain

$$\frac{d}{dt}(H_i^K + H_i^R) = \frac{d'}{dt}(H_i^K + H_i^R) + \epsilon_{ijk}\,\omega_j\,(H_k^K + H_k^R) = M_i \ . \qquad (66)$$

Here we have straightaway written the Eulerian form for a reference system that is rigidly fixed with respect to the enveloping body, i.e. a reference system that rotates with $\omega_i^K = \omega_i$. The component equations for concretely defined special cases can be derived from the general form (66).

When the axis of rotation of the rotor is also a principal axis of the enveloping body (the 3′-axis, for example), equation (66) can readily be expressed in terms of coordinates that are referred to the enveloping body. Indeed, with

$$\omega_1 = \omega_1^K = \omega_1^R \ ; \quad \omega_2 = \omega_2^K = \omega_2^R \ ; \quad \omega_3 = \omega_3^K \neq \omega_3^R \ ,$$
$$A^K + A^R = A \ ; \quad B^K + B^R = B^K + A^R = B \qquad (66a)$$

it follows from (66) that

$$(67)\qquad \begin{aligned} A\dot{\omega}_1 - (B-C^{K})\,\omega_2\omega_3 + C^{R}\omega_3^{R}\omega_2 &= M_1 , \\ B\dot{\omega}_2 - (C^{K}-A)\,\omega_3\omega_1 - C^{R}\omega_3^{R}\omega_1 &= M_2 , \\ C^{K}\dot{\omega}_3 - (A-B)\,\omega_1\omega_2 + C^{R}\dot{\omega}_3^{R} &= M_3 . \end{aligned}$$

Side by side with $\omega_1\ \omega_2\ \omega_3$, these equations also contain ω_3^R as an unknown. The necessary additional equation follows from the theorem of angular momentum applied to the rotor with respect to its axis of symmetry, i.e.

$$(68)\qquad C^{R}\dot{\omega}_3^{R} = M_3^{R} ,$$

where M_3^R represents the external torque that acts on the rotor in the direction of its own axis of symmetry. Here one has to distinguish the following three cases :

a) The rotor is supported within the enveloping body by means of frictionless bearings; the driving and the resisting torques can be neglected. In this case $M_3^R = 0$, and consequently

$$(69)\qquad \omega_3^{R} = \omega_{30}^{R} = \text{const.}$$

b) The angular velocity of the rotor relative to the enveloping body is kept constant by means of an controlled drive. In that case we have

$$(70)\qquad \omega_3^{R} = \omega_3 + \omega_{30}^{Z}$$

where ω_{3o}^Z is the constant relative angular velocity.

c) The angular velocity of the rotor relative to the enveloping body is a predetermined function of time (or of other variables of the system). In this case we have

$$\omega_3^R = \omega_3 + \omega_3^Z(t) \quad . \tag{71}$$

The function $\omega_3^Z(t)$ is generally chosen on the basis control considerations.

As regards Case a), equations (67) and (70) can be interpreted as the system of equations for an equivalent rigid gyro whose principal moments of inertia are A, B, C^K and whose self-excitation depends on the angular velocity. We obtain

$$\left.\begin{aligned} A\dot{\omega}_1 - (B-C^K)\,\omega_2\,\omega_3 &= M_1 - H^R\omega_2 \,, \\ B\dot{\omega}_2 - (C^K-A)\,\omega_3\,\omega_1 &= M_2 + H^R\omega_1 \,, \\ C^K\dot{\omega}_3 - (A-B)\,\omega_1\,\omega_2 &= M_3 \;. \end{aligned}\right| \tag{72}$$

where $H^R = C^R\omega_{3_0}^R$ is the constant component of the angular momentum of the rotor in the direction of the 3′-axis.

A completely analogous system of equations is obtained for Case b), the only difference being that C^K must be replaced by $C = C^K + C^R$ and H^R by $H^Z = C^R\omega_3^Z(t)$.

In Case c) we obtain a time-dependent angular momentum $H^Z(t)$, and in the third equation there also appears an additional excitation term $C^R\dot{\omega}_{3_0}^Z(t)$. This case is of importance when an attitude control of the enveloping body is to be effected by means of a control of the drive impressed upon the rotor it contains.

4.1.2. The Symmetrical Gyrostat without External Torques.

In this case we have $M_i = 0$ and $A^K = B^K$. Moreover, since $A^R = B^R$, it immediately follows that $A = B$, and the complete solution of the equations of motion can now be obtained quite readily. It follows from the third of equations (72) that $\omega_3 = \omega_{3o}$. If we now introduce the complex magnitude

$$\overset{\star}{\omega} = \omega_1 + i\omega_2 \tag{73}$$

the first two equations of (72) can be combined into

$$A\dot{\overset{\star}{\omega}} + i\left[(A-C^K)\,\omega_{3o} - H^R\right]\overset{\star}{\omega} = 0 \tag{74}$$

When the initial condition is defined by $\omega^\star(o) = \omega_o^\star$, the general solution of (74) takes the form

$$\overset{\star}{\omega} = \overset{\star}{\omega}_o\, e^{-i\nu t} \qquad \nu = \frac{1}{A}\left[(A-C^K)\,\omega_{3o} - H^R\right]. \tag{75}$$

The vector ω_i, which is made up of the components ω_{3o} and $\omega^\star$, generates a straight circular cone in the course of this motion (see Fig. 10), and its terminal point therefore rotates around the 3′-axis with an angular frequency ν. The angle of aperture of the polar cone is obtained from

$$\tan\mu = \frac{|\overset{\star}{\omega}_o|}{\omega_{3o}}.$$

Although the vector of the angular momentum H_i remains fixed with respect to space, with respect to the body it once again generates a straight circular cone, and this cone is co-axial with the polar cone. Since

$$H_i = (A\omega_1\ ,\ A\omega_2\ ,\ C^K\omega_{3o} + H^R) \tag{75a}$$

it follows that H_i must lie in the plane of ω_i and the 3′-axis. The angle of aperture ϑ of the angular momentum cone (fixed with respect to the body) is given by

$$\tan\vartheta = \tan\vartheta_0 = \frac{A|\overset{\star}{\omega}_0|}{C^K\omega_{30} + H^R} \quad . \tag{76}$$

The angle $\vartheta = \vartheta_0$ can also be interpreted as the Eulerian angle that describes the rotation of the 3′-axis (fixed with respect to the body) relative to the H_i-direction (3-axis), this latter being fixed with respect to space. The motion itself is a nutational motion in which the gyrostat's axis of rotation and axis of the figure (3′-axis) move around the direction of the angular momentum (which, of course, remains constant with respect to space). The azimuth speed $\dot{\psi}$ can be calculated from

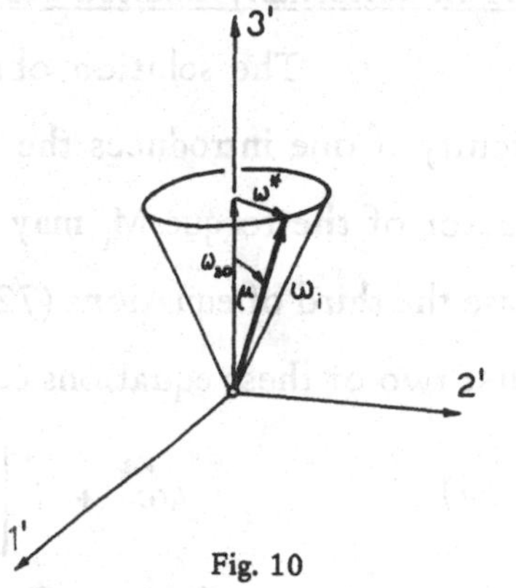

Fig. 10

$$|\overset{\star}{\omega}_0|^2 = \omega_1^2 + \omega_2^2 = \dot{\vartheta}^2 + (\dot{\psi}\sin\vartheta_0)^2 = \dot{\psi}^2\sin^2\vartheta_0 \tag{76a}$$

If this is combined with (76), one obtains

$$\dot{\psi} = \dot{\psi}_0' = \frac{|\overset{\star}{\omega}_0|}{\sin\vartheta_0} = \frac{1}{A}\sqrt{(A|\overset{\star}{\omega}_0|)^2 + (C^K\omega_{30} + H^R)^2} \quad . \tag{77}$$

This is a generalized expression for the speed of nutation. Indeed, if we put $H^R = 0$, it reduces to the known value for a single rigid body.

The Eulerian angle φ can be calculated from $\omega_{3_0} = \dot{\varphi} + \dot{\psi}\cos\vartheta$ or, alternately, it can be obtained from

$$\dot{\varphi} = \omega_{30} - \frac{1}{A}(C^K\omega_{30} + H^R) = \nu \tag{78}$$

Analogously to what we saw in the case of a single rigid body, it is possible to interpret the nutational motions of a symmetrical gyrostat as a relative rolling motion of polar cones and trace cones. Indeed, such an interpretation has been given by Leipholz [8]. However, such an interpretation can no longer be as readily visualized as it is in the case of an individual rigid body.

4.1.3. Externally Excited Motions of Symmetrical Gyrostats.

The solution obtained in 4.1.2 can be generalized without any difficulty if one introduces the condition $M^{\star} = M_1 + iM_2 \neq 0$. In other words, the vector of the torque M_i may be at a right angle to the axis of symmetry. In that case the third of equations (72) will once again yield $\omega_3 = \omega_{3_0}$, and with this the first two of these equations can be combined into the form

(79) $$A\dot{\omega}^{\star} + i\left[(A-C^{K})\,\omega_{30} - H^{R}\right]\omega^{\star} = M^{\star} .$$

If we now put $M^{\star}/A = m^{\star}$ and ν as given by (75), equation (79) becomes transformed into

(80) $$\dot{\omega}^{\star} + i\nu\omega^{\star} = m^{\star}$$

and this yields the general solution

(81) $$\omega^{\star} = e^{-i\nu t}\left[\omega^{\star}_{o} + \int_{o}^{t} m^{\star}(\tau)\, e^{i\nu\tau} d\tau\right] .$$

This solution can be explicitly calculated and discussed for many types of excitation functions $m^{\star}(t)$, the situation being similar to the case of the completely analogous solution (50) for a self-excited gyro.

Some further remarks should be made about the earlier Case c), this case being of importance for the attitude control of the enveloping body. If it is desired to perform an attitude control round all the three axes of the enveloping body, then it will be necessary for the rotor axis to be anchored inside the enveloping body in such a way as to enable it to be tilted. Alternately, it would be necessary to replace the single rotor by three separate rotors, for example, three rotors that have their axes in the directions of the principal axes of the enveloping body. Let C^{1R}, C^{2R}, C^{3R} represent the moments of inertia of the three rotors. If there are no external torques acting on the system, the system of equations (72) becomes replaced by

$$\left.\begin{aligned} A\dot{\omega}_1 - (B-C)\,\omega_2\omega_3 &= -C^{1R}\dot{\omega}^{1Z} + \omega_3 C^{2R}\omega^{2Z} - \omega_2 C^{3R}\omega^{3Z}, \\ B\dot{\omega}_2 - (C-A)\,\omega_3\omega_1 &= -C^{2R}\dot{\omega}^{2Z} + \omega_1 C^{3R}\omega^{3Z} - \omega_3 C^{1R}\omega^{1Z}, \\ C\dot{\omega}_3 - (A-B)\,\omega_1\omega_2 &= -C^{3R}\dot{\omega}^{3Z} + \omega_2 C^{1R}\omega^{1Z} - \omega_1 C^{2R}\omega^{2Z}. \end{aligned}\right| \qquad (82)$$

On the right-hand sides of these equations there now appear the functions ω^{1Z}, ω^{2Z}, ω^{3Z}, and these can be interpreted as control variables for the attitude control. By means of suitably chosen control equations they are brought into a relationship with the angles of the angular velocities that determine the position of the enveloping body. However, it is not proposed to examine this control problem here.

4.2. The Cardanic Suspended Gyro with a Symmetrical Rotor influenced by Gravity.

The cardanic suspended gyro, which is illustrated in Fig. 11, forms a system that consists of three bodies. Each of the three component bodies can rotate relative to the immediately adjacent body (or, in the case of the outermost

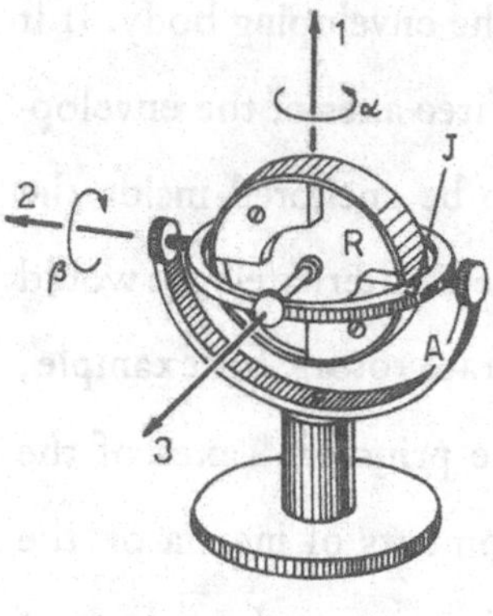

Fig. 11

body, relative to the frame), and in each case the rotation is performed round an axis that remains fixed with respect to the body itself. The considerations to be made about this system will be based on the following preliminary assumptions :

a) that the bearings of the gimbals and the rotor do not cause any friction resistance;

b) that the axes of the gimbals and the rotor intersect in a single point ;

c) that the gimbals and the rotor can be considered to be rigid ;

d) that the principal axes of the gimbals and the rotor can be brought into a normal position with respect to the frame, and that they coincide with the geometrical axes of the gimbals and the rotor.

The principal moments of inertia of the three component bodies will be represented as follows :

$$\begin{array}{llll} \text{Those of the rotor :} & A^R & B^R & C^R \\ \text{Those of the inner gimbal :} & A^J & B^J & C^J \\ \text{Those of the outer gimbal :} & A^A & - & - \end{array} \qquad (83)$$

The position in space of all the three component bodies will be described by means of the cardanic (gimbal) angles $\alpha\ \beta\ \gamma$, these angles corresponding respectively to rotations about the axis of the outer gimbal, about the axis of the inner gimbal, and about the axis of the rotor. The components of the angular velocities of the three component bodies $\omega_i^R\ \ \omega_i^J\ \ \omega_i^A$ are then given by

$$\left.\begin{aligned} \omega_1^R &= \dot{\alpha}\cos\beta\cos\gamma + \dot{\beta}\sin\gamma \ , \\ \omega_2^R &= -\dot{\alpha}\cos\beta\sin\gamma + \dot{\beta}\cos\gamma \ , \\ \omega_3^R &= \dot{\gamma} + \dot{\alpha}\sin\beta \ , \end{aligned}\right| \qquad (84)$$

$$\left.\begin{aligned} \omega_1^J &= \dot{\alpha}\cos\beta & \omega_1^A &= \dot{\alpha} \ , \\ \omega_2^J &= \dot{\beta} \ , & \omega_2^A &= 0 \ , \\ \omega_3^J &= \dot{\alpha}\sin\beta \ , & \omega_3^A &= 0 \ . \end{aligned}\right| \qquad (84)$$

the definition in each case being referred to the coordinate system that remains fixed with respect to the body to which the components relate.

The methods to be employed for the purposes of calculating the behaviour of the gyro can be chosen in accordance with the particular case that is being treated. Thus, when calculating the behaviour of a cardanic suspended gyro with a symmetrical rotor influenced by gravity, it will be found advantageous to make use of energy expressions and then to formulate the equations of motion in accordance with the Lagrangian method. When calculating the behaviour of a cardanic suspended gyro with an unsymmetrical rotor, on the other hand, the Eulerian form of the equations of motion presents advantages.

4.2.1. The General Solution.

In addition to the assumptions that have already been mentioned, it will now be supposed that we have a symmetrical rotor (i.e. that $A^R = B^R$), that the axis of the outer gimbal (1-axis) is in a vertical position, and that the position of the joint centre of gravity of the rotor and the inner gimbal is located on the axis of symmetry of the rotor (i.e. $s_1 = s_2 = 0$, and $s_3 = s \neq 0$). In this case, putting $s > 0$ and $\beta = + \pi/2$, one obtains a description of an statically unstable system (upright gyro) ; likewise, with $\beta = -\pi/2$, one obtains a description of a statically stable system (hanging gyro).

For the purpose of formulating the equations of motion we need the expressions for the kinetic energy and the potential energy of the complete system. Since the axis of the outer gimbal is in a vertical position, the potential energy U depends only on the angle β, and is therefore independent of α. Using G to represent the weight of the rotor and the inner gimbal, one thus obtains

$$U = Gs \sin\beta \; . \tag{85}$$

For the kinetic energy one can write

$$T = T^A + T^J + T^R = \sum_{A,J,R,} \frac{1}{2} \Theta_{ij} \omega_i \omega_j =$$

$$= \frac{1}{2} \left[A^A \omega_1^{A^2} + (A^J \omega_1^{J^2} + B^J \omega_2^{J^2} + C^J \omega_3^{J^2}) + (A^R \omega_1^{R^2} + B^R \omega_2^{R^2} + C^R \omega_3^{R^2}) \right]$$

and, after making the appropriate substitutions and re-arrangements, on thus obtains

$$T = \frac{1}{2} \left\{ \dot{\alpha}^2 [A\cos^2\beta + (A^A + C^J) \sin^2\beta] + \dot{\beta}^2 B + C^R (\dot{\gamma} + \dot{\alpha} \sin\beta)^2 \right\} . \tag{86}$$

where A and B represent the abbreviations

$$A = A^R + A^J + A^A \; ; \quad B = B^R + B^J \; .$$

Analogously to the case of the Lagrangian gyro, it is now possible to find three integrals of the equations of motion. Two of these follow from the fact that both γ and α are cyclical coordinates : they do not appear as such in the expressions for T and U, but only in the form of their derivatives. If we consider the Lagrangian equation for γ together with equation (86), it follows that

$$\frac{\partial T}{\partial \dot{\gamma}} = C^R (\dot{\gamma} + \dot{\alpha} \sin\beta) = \text{const}$$

or

$$\omega_3^R = \dot{\gamma} + \dot{\alpha} \sin\beta = \omega_0 \quad . \tag{87}$$

It can therefore be seen that the component of the angular momentum along the axis of the figure of the rotor, and consequently also the angular velocity, remain constant. In view of the assumption that no friction losses occur in the bearings, this has a readily understandable physical meaning.

In a corresponding manner, it follows from the Lagrangian equation for α that the vertical component of the angular momentum of the overall system must also remain constant, i.e.

$$\frac{\partial T}{\partial \dot{\alpha}} = \dot{\alpha} \left[A\cos^2\beta + (A^A + C^J) \sin^2\beta \right] + C^R \omega_0 \sin\beta = H_0 = \text{const.} \tag{88}$$

We can again find a readily visualizable explanation for this fact : the only external torques that can act on the overall system are gravity torques and such other torques as the bearing of the outer gimbal transfers onto the gimbal axis. Since this bearing is in a vertical position and has been assumed to cause no friction losses, the vectors of both types of torques must always lie in the horizontal plane.

Apart from the two integrals of the angular momentum, i.e. (87) and

(88), we can also apply the energy theorem to the cardanic suspended gyro. We therefore have

$$T + U = E_0 \quad ,$$

$$(89) \quad \dot{\alpha}^2 \left[A\cos^2\beta + (A^A + C^J)\sin^2\beta \right] + \dot{\beta}^2 B + C^R\omega_0^2 + 2Gs\sin\beta = 2E_0 .$$

Using these three integrals, we can then obtain the complete solution as follows : from (88) we find that

$$(90) \qquad \dot{\alpha} = \frac{H_0 - C^R\omega_0\sin\beta}{A\cos^2\beta + (A^A + C^J)\sin^2\beta}$$

and then substitute this in equation (89). If we now solve this equation for $\dot{\beta}$, we find that

$$(91) \quad B\dot{\beta}^2 = 2E_0 - C^R\omega_0^2 - 2Gs\sin - \frac{(H_0 - C^R\omega_0\sin\beta)^2}{A\cos^2\beta + (A^A + C^J)\sin^2\beta}$$

or, more generally,

$$(91a) \qquad \dot{\beta}^2 = F(\beta) \ .$$

If we now introduce

$$(92) \qquad \sin\beta = u \ ; \ \dot{\beta} = \frac{\dot{u}}{\cos\beta}$$

into equation (91), we obtain

$$(93) \quad \dot{u}^2 = (1-u^2)\left\{ \frac{2E_0 - C^R\omega_0^2}{B} - \frac{2Gs}{B}u - \frac{(H_0 - C^R\omega_0 u)^2}{B\,[A-(A^J + A^A - C^J)\,u^2]} \right\} = U(u).$$

As the moments of inertia of the gimbals tend towards zero, this equation will assume the well-known form that applies in the case of the Lagrangian gyro. However, the gyro function U(u) in this equation is no longer a simple polynomial, but rather a rational fraction function of u, and its integration does not lead to any known functions. Nevertheless, the integration can always be performed by numerical or graphical means. The integration of equation (91a) yields

$$t = t_0 + \int \frac{d\beta}{\sqrt{F(\beta)}} = t(\beta) \ , \tag{94}$$

and from this it follows that $\beta = \beta(t)$. If this value is substituted in equation (90), a further integration yields $\alpha = \alpha(t)$. Substituting this in equation (87) and integrating once again, we obtain $\gamma = \gamma(t)$. With this we have found the complete solution for the motions of a cardanic suspended gyro whose outer gimbal rotates round a vertical axis and whose rotor is influenced by gravity.

4.2.2. The Stability of the Cardanic Suspended Gyro when the Rotor Axis is vertical.

Of all the possible solutions it is here proposed to examine in greater detail only one special motion of the cardanic suspended gyro, i.e. the one that occurs when the rotor axis is in a vertical position. Our starting point for this examination is the Lagrangian equation for the gimbal angle β, an equation that has not yet been formulated. From

$$\frac{d}{dt}\left(\frac{\partial T}{\partial \dot{\beta}}\right) - \frac{\partial T}{\partial \beta} + \frac{\partial U}{\partial \beta} = 0$$

and equations (85) and (86) we obtain the equation of motion

$$(95)\quad B\ddot{\beta} + (A^R + A^J - C^J)\sin\beta\cos\beta\,\dot{\alpha}^2 - C^R\omega_0\cos\beta\,\dot{\alpha} + Gs\cos\beta = 0 \,.$$

This equation is satisfied when

$$(96)\qquad \cos\beta = 0 \text{ for } \beta = \pm\frac{\pi}{2} \pm n\pi \,,\ \dot{\beta} = \ddot{\beta} = 0.$$

If we now substitute this in equation (90), we obtain

$$(97)\qquad \dot{\alpha} = \frac{H_0 \pm C^R\omega_0}{A^A + C^J} = \text{const.}$$

It can thus be seen that it is possible for the gyro to rotate with its rotor axis in a vertical position, that is to say, pointing in the same direction as the axis of the outer gimbal; in this case the two gimbals will rotate with a constant angular velocity $\dot{\alpha}$ around the axis of the outer gimbal.

This particular motion is not by any means always stable. This will be realized most readily if one examines motions that are very close to $\beta_0 = \frac{\pi}{2}$ for example, by putting $\beta = \frac{\pi}{2} + \vartheta$, where $\vartheta \ll 1$. It then follows from (96) that

$$(98)\qquad B\ddot{\vartheta} + \left[-\dot{\alpha}^2(A^R + A^J - C^J) + \dot{\alpha}C^R\omega_0 - Gs\right]\vartheta = 0 \,.$$

The expression contained in the square bracket can be interpreted as a restoring force function $r(\dot{\alpha})$. Only positive values of r will lead to stable solutions for $\vartheta(t)$. Examining the functions $r(\dot{\alpha})$, which are plotted in Fig. 12, it can be seen that stable motions become possible only when

$$\dot{\alpha}_1 > \dot{\alpha} > \dot{\alpha}_2 \tag{99}$$

In the case of an upright gyro with $s > 0$ we have $0 < \dot{\alpha}_1 < \dot{\alpha}_2$ whereas in the case of a gyro with $s < 0$ we have $\dot{\alpha}_2 < 0 < \dot{\alpha}_1$. From this it follows, among others, that an upright gyro can never be stable if the gimbals remain at rest, and this no matter how fast the rotor may rotate. If stability is to be obtained, it will be necessary to impress upon the gimbal an angular velocity $\dot{\alpha}$ (in the same sense as the rotation of the rotor), and $\dot{\alpha}$ must be such as to satisfy (99). The hanging gyro, on the other hand, can be stable when $\dot{\alpha} = 0$ indeed, it can even be stable when the gimbal is rotating at moderate speed in the opposite direction to the rotation of the rotor. As regards the limiting values of $\dot{\alpha}$ it follows from $r(\dot{\alpha})=0$ that

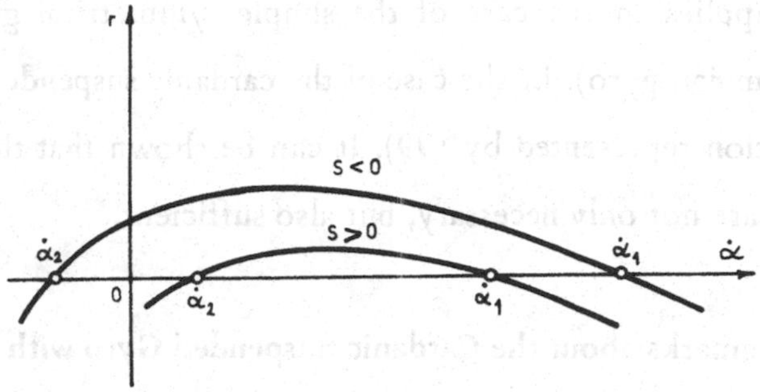

Fig. 12

$$\left.\begin{matrix}\dot{\alpha}_1\\ \dot{\alpha}_2\end{matrix}\right\} = \frac{C^R\omega_0}{2(A^R+A^J-C^J)}\left[1 \pm \sqrt{1 - \frac{4Gs(A^R+A^J-C^J)}{(C^R\omega_0)^2}}\right]. \tag{100}$$

From this it immediately follows that real values of $\dot{\alpha}$ can only exist when the condition

$$(C^R\omega_0)^2 \geqslant 4Gs(A^R+A^J-C^J) \tag{101}$$

is satistfied. It can be seen that this is a generalized form of the well-known stability condition

$$(C^R \omega_0)^2 \geqslant 4GsA^R \quad (101a)$$

that applies in the case of the simple symmetrical gyro influenced by gravity (Lagrangian gyro). In the case of the cardanic suspended gyro we have the further condition represented by (99). It can be shown that the two conditions (99) and (101) are not only necessary, but also sufficient.

4.3. Remarks about the Cardanic Suspended Gyro with an Unsymmetrical Rotor

The stability behaviour of a cardanic suspended gyro with an unsymmetrical rotor differs in a remarkable manner from the well-known behaviour of a simple gyro on which no external torques are acting (Eulerian gyro). The rotations of an Eulerian gyro are always stable around the axes of the largest and the smallest of the principal moments of inertia, and always unstable around the axis of the middle moment of inertia. It is quite true that also in the case of the cardanic suspended gyro it is possible to have permanent rotations about coincident principal axes of the three component bodies. However, the stability of these rotations does not only depend on the rotor, but also on the mass ratios of the two gymbals. If three rotations defined by Prandtl are chosen as reference rotations about the three principal axes of the rotor,then, in addition to the cases that are known by virtue of the analysis of the Eulerian gyro, there exist also particular mass distributions for which either all three rotations remain stable or,alternately, only one of the rotations remains stable, while the other two are unstable.

In the case of the cardanic suspended gyro, with the axes of the system at right angles to each other, the three Prandtl rotations can be realized as follows:

1) A rotation about the axis of the rotor with the gimbals remaining in rest;

2) A rotation of the system about the axis of the outer gimbal with the axis of the largest principal moment of inertia of the rotor remaining parallel to the axis of the outer gimbal (that is to say, the rotor does not perform any rotations with respect to the inner gimbal);

3) As in 2), but with the rotor displaced through 90°;

Without examining the theory itself, which is explained in detail in another of the author's publications [9], we shall here limit ourselves to reporting the results. In this connection we shall assume that changes in the mass distribution can be brought about by fixing additional weights at the points of the inner gimbal where the rotor bearings are situated. If the increase in the moments of inertia of the inner gimbal caused by these additional weights is represented by θ, i.e.

$$A^J = A_0^J + \theta \quad ; \quad B^J = B_0^J + \theta \;, \tag{102}$$

then the stability expression for the three Prandtl rotations can be regarded as functions of θ. In a theory of the first order one obtains, as necessary conditions for stable rotations, inequalities of the form $S_i > 0 (i = 1, 2, 3,)$, where

$$\left.\begin{aligned} S_1 &= \left[A^R - C^R + \frac{1}{2}(A_0^J + B_0^J + A^A) + \theta\right]\left[B^R - C^R + \frac{1}{2}(A_0^J + B_0^J + A^A) + \theta\right], \\ S_2 &= (A^R - B^R)(A^R - C^R + A_0^J - C^J + \theta)\;, \\ S_3 &= (B^R - A^R)(B^R - C^R + A_0^J - C^J + \theta)\;. \end{aligned}\right| \tag{103}$$

The functions S (θ) are plotted in Fig.13. The points of intersection of these curves with the abscissae, i.e. θ_1 to $\overline{\theta_4}$ can be read directly from (103). They subdivide the θ-domain into five partial domains, and each of these corresponds to a different stability behaviour. As regards the cardanic suspended gyro on which Fig. 13 is based, this can best be seen from the following table:

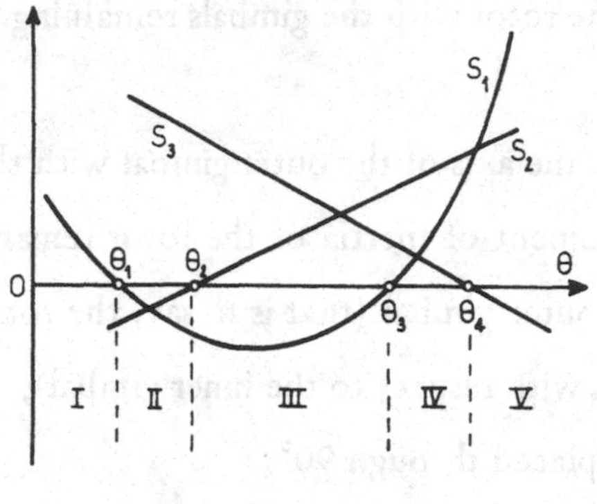

Fig. 13

Prandtl rotation	I $\theta<\theta_1$	II $\theta_1<\theta<\theta_2$	III $\theta_2<\theta<\theta_3$	IV $\theta_3<\theta<\theta_4$	V $\theta_4<\theta$
No. 1 $S_1 > 0$	+	–	–	+	+
No. 2 $S_2 > 0$	+	+	+	+	–
No. 3 $S_3 > 0$	–	–	+	+	+

In this table a plus sign (+) indicates that the appropriate stability condition is complied with, while a minus sign (–) denotes instability. It can be seen that two of the rotations in each of Domains I, III and IV are stable, while the third rotation about the axis of the "middle" moment of inertia is unstable. This corresponds to the stability behaviour of the Eulerian gyro. On the other hand, the motion in Domains II and IV deviates from the well-known behaviour of the single rigid body; when $\overline{\theta_1 < \theta < \theta_2}$ (i.e. in Domain II), there is only one stable Prandtl rotation, while there are three such rotations when $\theta_3 < \theta < \theta_4$ (i.e. in Domain V). This result, moreover, has been experimentally confirmed; it can thus be seen that the picture of the necessary stability conditions derived from a theory of the first order of approximation effectively corresponds to reality.

5. Gyro Systems

5.1. Equations of Motion

Just as in the case of a single rigid body, two types of equations of motion can be distinguished for the case of gyro systems, i.e. when we have a system of several rigid bodies that are coupled to each other. These two types can be respectively described as Lagrangian and Eulerian. Energy expressions are used as the starting point for formulating equations of the first type, while equations of the Eulerian type are generally derived by applying the theorem of angular momentum to the component bodies of the system and then, by transformation, to the overall system. Without entering into the details of the theory, we shall here limit ourselves to discussing a few important propositions and results.

5.1.1. Equations of Motion of the Lagrangian Type.

Theories based on Lagrange start with a statement of the kinetic energy of the system, i.e.

$$T = \frac{1}{2} \int_m \dot{x}_i \dot{x}_i \, dm \qquad (104)$$

Using the relationship that exists between the cartesian coordinates x_i and the generalized coordinates q_ξ ,which may, for example, be of the holonomous form

$$x_i = x_i\,(q_1 \ldots q_n) \tag{105}$$

since

$$x_i = \frac{\partial x_i}{\partial q_\xi}\, q_\xi \qquad (\xi = 1, \ldots, n) \tag{106}$$

this expression for the kinetic energy is then brought into the form of the double sum

$$T = \frac{1}{2}\, a_{\xi\eta}\, \dot{q}_\xi\, \dot{q}_\eta \tag{107}$$

In this expression the terms

$$a_{\xi\eta} = a_{\eta\xi} = \int_m \frac{\partial x_i}{\partial q_\xi}\, \frac{\partial x_i}{\partial q_\eta}\, dm \tag{108}$$

represent the generalized masses. For the purposes of a further transformation of this expression a distinction has to be made between the cyclical and non-cyclical coordinates q_ξ . Let $q_1, \ldots, q_m$ be the non-cyclical coordinates; they will be characterized by means of the suffices α , β , γ , δ and let $q_{m+1}, \ldots, q_n$ be the cyclical coordinates ; they will be characterized by means of the suffices κ, λ, μ, ν.

By means of an appropriate separation of the mass matrix

$$a_{\xi\eta} = \left|\begin{array}{c|c} a_{\alpha\beta} & a_{\alpha\kappa} \\ \hline a_{\alpha\kappa} & a_{\kappa\lambda} \end{array}\right| \tag{108a}$$

(107) can be brought into the form

$$T = \frac{1}{2} a_{\alpha\beta} \dot{q}_\alpha \dot{q}_\beta + a_{\alpha\kappa} \dot{q}_\alpha \dot{q}_\kappa + \frac{1}{2} a_{\kappa\lambda} \dot{q}_\kappa \dot{q}_\lambda \tag{109}$$

In accordance with our definition of the cyclical coordinates, we here have

$$\frac{\partial T}{\partial \dot{q}_\mu} = a_{\alpha\mu} \dot{q}_\alpha + a_{\mu\lambda} \dot{q}_\lambda = p_\mu , \qquad (\mu = m+1, \ldots, n) \tag{110}$$

where p_μ are the constant generalized momentums. Since the matrix $a_{\mu\lambda}$ is non-singular – otherwise the kinetic energy, which is made up solely of the cyclic coordinates, would not be a positively definite quadratic form – it is possible to calculate from (110) that

$$\dot{q}_\lambda = a^{-1}_{\mu\lambda}(p_\mu - a_{\alpha\mu} \dot{q}_\alpha) , \qquad (\lambda = m+1, \ldots, n) \tag{111}$$

By substituting this in (109) it becomes possible to eliminate the cyclical coordinates. At this point it is usual to introduce the Routhian function

$$R = T - p_\kappa \dot{q}_\kappa \tag{112}$$

to replace the kinetic energy T, and one can then obtain the system

$$\frac{d}{dt}\left(\frac{\partial R}{\partial \dot{q}_\alpha}\right) - \frac{\partial R}{\partial q_\alpha} = Q_\alpha \ , \qquad (\alpha = 1\ , \ \ldots, \ m) \tag{113}$$

from the Lagrangian equations of motion of the second type. Only the non-cyclical coordinates appear in this system, coordinates that are generally externally observable. Together with the n-m relationships of (110), one now has at one's disposal n equations for the determination of the n generalized coordinates.

5.1.2. The Equations of Motion according to Kelvin and Tait.

A more detailed examination of the Routhian function R and its properties makes it possible to bring the equations of motion (113) into a form that lends itself more readily to interpretation. Substituting (111) in (109) and (112), one first of all obtains the Routhian function

$$\begin{aligned} R = {} & \frac{1}{2} a_{\alpha\beta}\, \dot{q}_\alpha \dot{q}_\beta + a_{\alpha\kappa}\, \dot{q}_\alpha\, a^{-1}_{\mu\kappa} (p_\mu - a_{\beta\mu} \dot{q}_\beta) \ + \\ & + \frac{1}{2} a_{\kappa\lambda}\, a^{-1}_{\mu\kappa} (p_\mu - a_{\alpha\mu} \dot{q}_\alpha)\, a^{-1}_{\nu\lambda} (p_\nu - a_{\beta\nu} \dot{q}_\beta) \ - \\ & - p_\kappa\, a^{-1}_{\mu\kappa} (p_\mu - a_{\alpha\mu} \dot{q}_\alpha)\ . \end{aligned} \tag{114}$$

When this is multiplied out, one notes that R consists of three components that are respectively quadratic, linear and constant with respect to the non-cyclic speeds, i.e.

$$R = R_2 + R_1 - R_0 .$$

Taking account of the rules that govern the multiplication of matrices, these three components can be written as follows:

$$R_2 = \frac{1}{2}\left(a_{\alpha\beta} - a^{-1}_{\kappa\lambda}\, a_{\alpha\kappa}\, a_{\beta\lambda}\right)\dot{q}_\alpha \dot{q}_\beta ,$$

$$R_1 = a^{-1}_{\kappa\lambda}\, a_{\kappa\alpha}\, p_\lambda\, \dot{q}_\alpha \quad , \qquad (115)$$

$$R_0 = \frac{1}{2}\, a^{-1}_{\kappa\lambda}\, p_\kappa p_\lambda \; .$$

Substituting this in (113), one now obtains the equations of motion in the following form

$$\frac{d}{dt}\left(\frac{\partial R_2}{\partial \dot{q}_\alpha}\right) - \frac{\partial R_2}{\partial q_\alpha} = Q_\alpha - \frac{d}{dt}\left(\frac{\partial R_1}{\partial \dot{q}_\alpha}\right) + \frac{\partial R_1}{\partial q_\alpha} - \frac{\partial R_0}{\partial q_\alpha} \; . \qquad (116)$$

The two central terms on the right -hand side of this equation can be combined as follows:

$$\frac{\partial R_1}{\partial q_\alpha} - \frac{d}{dt}\left(\frac{\partial R_1}{\partial \dot{q}_\alpha}\right) = \left[\frac{\partial}{\partial q_\alpha}\left(a^{-1}_{\kappa\lambda}\, a_{\kappa\beta}\, p_\lambda\right) - \frac{\partial}{\partial q_\beta}\left(a^{-1}_{\kappa\lambda}\, a_{\kappa\alpha}\, p_\lambda\right)\right]\dot{q}_\beta =$$

$$= g_{\alpha\beta}\, \dot{q}_\beta = G_\alpha \; . \qquad (117)$$

where

$$(118)\qquad g_{\alpha\beta} = \frac{\partial}{\partial q_\alpha}(a^{-1}_{\kappa\lambda}\, a_{\kappa\beta}\, p_\lambda) - \frac{\partial}{\partial q_\beta}(a^{-1}_{\kappa\lambda}\, a_{\kappa\alpha}\, p_\lambda) = - g_{\beta\alpha} \,.$$

The terms G_a are known as generalized gyroscopic forces. In accordance with (117), they depend in a linear manner on the changes of the non-eliminated coordinates, the matrix $g_{\alpha\beta}$ of the gyroscopic terms having a skew symmetry. It therefore follows that $g_{aa} = 0$. The equations of motion (116) can now be written in the form in which they were stated by Kelvin and Tait, i.e.

$$(119)\qquad \frac{d}{dt}\left(\frac{\partial R_2}{\partial \dot{q}_a}\right) - \frac{\partial R_2}{\partial q_a} = Q_a + G_a - \frac{\partial R_0}{\partial q_a}\,, \qquad (\alpha = 1, \ldots, m)\,.$$

These equations show that the motions of a system can be calculated in such a way as if only the non-cyclical positional coordinates were present, that is to say, coordinates that for the most part can be observed from outside the system. In that case, however, one has to effect the following three changes as compared with the usual Lagrangian equations:

1. The generalized gyroscopic forces G_a have to be added to the generalized external forces Q_a. We shall subsequently discuss these generalized gyroscopic forces in greater detail.
2. The kinetic constraints $-\partial R_0 / \partial q_\alpha$ have to be added to the external forces. Such kinetic restraints, for example, can be represented by centrifugal forces that come into being when cyclical coordinates are present. The magnitude R_0 itself can be interpreted as the potential of these forces.
3. The magnitude R_2 (as defined in (115)) must be introduced to replace the kinetic energy. Although R_2 is once again a quadratic function of the

$\dot{q}_a$, it is characterized by a changed mass matrix. In practice this is equivalent to a change in the moments of inertia.

A system with $G_\alpha = 0$ is described as a gyroscopically uncoupled system. It can readily be seen from (117) that this condition is fulfilled, for example, when $a_{\kappa a} = 0$. Considering (115) together with (110), it can be seen that in this case one obtains

$$R_0 = \frac{1}{2} a_{\kappa\lambda} \dot{q}_\kappa \dot{q}_\lambda \;;\quad R_1 = 0 \;;\quad R_2 = \frac{1}{2} a_{\alpha\beta} \dot{q}_\alpha \dot{q}_\beta . \qquad (119a)$$

R_2 is now the real kinetic energy of the visible and generally non-cyclical positional coordinates, while R_0 is the kinetic energy of the hidden cyclical coordinates. The equations of motion (119) assume the Lagrangian form, but on the right-hand sides of these equations one now has to add the conservative forces of the kinetic constraints, forces that can be derived from the potential R_0.

It should further be pointed out that equations of motion of a similar kind can also be formulated when the relationships between the coordinates are rheonomous rather than holonomous, that is to say, when the equations (105) defining the relationship contain not only the derivatives, but also time.

It is a characteristic feature of the generalized gyroscopic forces defined by (117) that they do not perform any work during the motions of the system. These forces always obey the equation

$$dA = G_a dq_a = G_a \dot{q}_a dt = g_{\alpha\beta} \dot{q}_a \dot{q}_\beta dt = \frac{1}{2} (g_{\alpha\beta} + g_{\beta a}) \dot{q}_a \dot{q}_\beta dt = 0 . \qquad (119b)$$

Since $g_{\alpha\beta} = - g_{\beta\alpha}$, the term contained in the bracket disappears. Viceversa, the skew symmetry of the matrix $g_{\alpha\beta}$ is a direct consequence of the requirement that these forces should be such as to perform no work. It is therefore possible to define the gyroscopic forces as forces that depend in a linear manner on the derivatives and for which the work remains zero no matter what motions the system may perform. Of course, such a general definition will also cover forces that are not brought into being by built in gyros or flywheels. Thus, to give but one example, in electrical networks it is possible to realize gyroscopic terms with the help of a purely electrical construction element, i.e. the gyrator.

The following important relationships follow from the skew symmetry of the matrix $g_{\alpha\beta}$

$$\left.\begin{aligned} \det(g_{\alpha\beta}) &= 0 \quad \text{for odd } m \\ \det(g_{\alpha\beta}) &> 0 \quad \text{for even } m \end{aligned}\right\} \quad \alpha,\beta = (1,\dots,m) \tag{120}$$

5.1.3. Equations of Motion of the Eulerian Type.

When this type of equation is applied to systems consisting of several bodies, a number of typical difficulties appear that do not have to be faced in the well-known case of the single rigid body. These difficulties may be summarized as follows:

1) The reference systems fixed with respect to the component bodies rotate with respect to each other
2) The origins of these reference systems can also move with respect to each other
3) The forces and torques of the reactions between the component bodies appear as additional unknowns in the equations of motion.

In most cases it will therefore be advantageous to refer all motions to a suitably chosen reference system; such a reference system, for example, may be the one that is constituted by the principal axes of the component body that is of particularly immediate interest. If this system rotates with an angular velocity Ω_i, the theorem of angular momentum can be written in the completely general form

$$\dot{H}_i = \mathring{H}_i + \epsilon_{ijk}\Omega_j H_k = M_i \tag{121}$$

In this expression the term $\mathring{H}_i$ designates the derivative with respect to time of the overall angular momentum of the system H_i as obtained in the reference system. The displacements and rotations of the component bodies with respect to each other make the calculation of H_i a particularly toilsome process, and this all the more so as one also has to take into account the translational momentums of the component bodies. Bearing in mind equations (31) and (24) and considering just one of the component bodies, whose reference point P is fixed with respect to the body and moves with a velocity v_i^P, one can first of all obtain the Eulerian Equations

$$\theta^P_{ij}\overset{\star}{\omega}_j + \epsilon_{ijk}\omega_j\theta^P_{kl}\omega_l + m\epsilon_{ijk}r_j\dot{v}^P_k = M^P_i \ . \tag{122}$$

where the term $\overset{\star}{\omega}_j$ denotes the derivative with respect to time as obtained in the reference system fixed with respect to the body. The motions of the component bodies are now judged from the reference system that rotates with an angular velocity Ω_i. If the angular velocity of the α-th component body relative to this reference system is denoted by ω_i^a then one obtains

$$\omega_i = \Omega_i + \omega_i^a ,$$

and the following relationships apply with regard to the derivatives in the various reference systems:

$$\begin{aligned}
\dot{\Omega}_i &= \mathring{\Omega}_i = \overset{\star}{\Omega}_i + \epsilon_{ijk}\, \omega^{\alpha}_j \Omega_k \,, \\
\dot{\omega}^{\alpha}_i &= \mathring{\omega}^{\alpha}_i + \epsilon_{ijk}\, \Omega_j \omega^{\alpha}_k = \overset{\star}{\omega}^{\alpha}_i + \epsilon_{ijk}\, \Omega_j \omega^{\alpha}_k \,, \\
\overset{\star}{\omega}_i &= \overset{\star}{\Omega}_i + \overset{\star}{\omega}^{\alpha}_i = \mathring{\Omega}_i + \mathring{\omega}^{\alpha}_i + \epsilon_{ijk}\, \Omega_j \omega^{\alpha}_k \,.
\end{aligned} \tag{123}$$

With the help of the above equation (122) can now be transformed into

$$\begin{aligned}
&\theta^{P\alpha}_{ij}\,(\mathring{\Omega}_j + \mathring{\omega}^{\alpha}_j + \epsilon_{jkl}\, \Omega_k \omega^{\alpha}_l) + \epsilon_{ijk}\,(\Omega_j + \omega^{\alpha}_j)\, \theta^{P\alpha}_{kl}\,(\Omega_l + \omega^{\alpha}_l) + \\
&\qquad + m^{\alpha}\, \epsilon_{ijk}\, r^{\alpha}_j\, \dot{v}^{P\alpha}_k = M^{P\alpha}_i + M^{B\alpha}_i
\end{aligned} \tag{124}$$

where $M^{B\alpha}_i$ are the torques that the reference body exerts on the α–th component body, while $M^{P\alpha}_i$ represents the remaining torques that are exerted on the α–th component body. In addition to equation (124), one also has to apply the theorem of momentum to the α–th component body. This theorem can be written in the form

$$m^{\alpha}\, \dot{v}^{S\alpha}_i = m^{\alpha}\,(\mathring{v}^{S\alpha}_i + \epsilon_{ijk}\, \Omega_j v^{S\alpha}_k) = F^{\alpha}_i + F^{B\alpha}_i \tag{125}$$

Equations (124) and (125) can be used, for example, whenever one wants to judge the motions of an individual gyro from a moving reference system.

By means of the elimination of the reaction forces and torques, in most cases these are not of any immediate interest, it now becomes possible to combine the equations for the individual bodies. However, the equations obtained in this manner do not lend themselves very readily to interpretation, and in most cases where it is desired to solve practical problems they have to be brought into a form that lends itself to treatment by means of electronic computers. Suitable algorithms for this purpose have been stated by Wittenburg [10], among others. Lurge [11] and Roberson [12] have also made some substantial contributions.

5.2. Approximations

Two methods have shown themselves to the particularly fruitful for the purposes of an appropriate determination of the behaviour of gyro systems : firstly, the method of small oscillations and, further, a limitation that restricts the results to systems with fast-running gyros.

5.2.1. Small Oscillations of Gyro Systems

The method of small oscillations leads to a linearization of the equations of motion. In this method the deviations of the system coordinates from known values of functions of time are regarded as small magnitudes. If $q_{\alpha o}(t)$ represents a known solution of motion, then one assumes that the immediately adjacent motions are described by

$$q_\alpha(t) = q_{\alpha o}(t) + x_\alpha(t) \ , \quad (\alpha = 1, \ldots, m) \tag{126}$$

The carrying out of the process of linearization, in which all the magnitudes that are dependent on q_α (such as the generalized masses and the generalized forces) have to be developed into Taylor series of the x_α terms, will in all cases lead to equations of motion that have the form

$$a_{\alpha\beta}\ddot{x}_\beta + b_{\alpha\beta}\dot{x}_\beta + c_{\alpha\beta}x_\beta = P_\alpha \ , \quad (\alpha = 1, \ldots, m) \tag{127}$$

The terms of the second and higher order in the x_β terms are combined in the expression P_α; but these terms can also contain components that are functions of time. The P_α terms are neglected when small oscillations in autonomous systems are being examined.

As regards the coefficients that occur in equation (127), $a_{\alpha\beta} = a_{\beta\alpha}$ is always symmetrical. The quadratic form

(128) $$T = \frac{1}{2} a_{\alpha\beta} \dot{x}_\alpha \dot{x}_\beta$$

is always positive definite; it constitutes the component of the kinetic energy that is made up of the non-cyclical derivatives .

The matrix $b_{\alpha\beta}$ of (127) can contain damping, exciting and gyroscopic terms, and also other terms that are brought about by changes in the masses and the moments of inertia. In rocket technology, for example, these latter components are known as jet damping. Lastly, the matrix $c_{\alpha\beta}$ can contain components that are brought into being by static or kinetic constraints (by centrifugal forces, for example) or even by changes in the damping coefficients. It will be advantageous to represent each of the matrices $b_{\alpha\beta}$ and $c_{\alpha\beta}$ as the sum of two matrices, of which one is symmetrical and the other has a skew symmetry. One then has

(129) $$b_{\alpha\beta} = d_{\alpha\beta} + g_{\alpha\beta} \quad \text{with} \quad \begin{cases} d_{\alpha\beta} = d_{\beta\alpha}, \\ g_{\alpha\beta} = -g_{\beta\alpha}, \end{cases}$$
$$c_{\alpha\beta} = f_{\alpha\beta} + e_{\alpha\beta} \quad \text{with} \quad \begin{cases} f_{\alpha\beta} = f_{\beta\alpha}, \\ e_{\alpha\beta} = -e_{\beta\alpha}. \end{cases}$$

Introducing (129) into the system of the equations of motion, this latter assumes the form

(130) $$a_{\alpha\beta} \ddot{x}_\beta + d_{\alpha\beta} \dot{x}_\beta + g_{\alpha\beta} \dot{x}_\beta + f_{\alpha\beta} x_\beta + e_{\alpha\beta} x_\beta = 0 \ .$$

$$(\alpha, \beta = 1,2,\ldots, m)$$

All the terms that occur in this system can be interpreted as generalized forces. Thus

$A_\alpha = - a_{\alpha\beta}\ddot{x}_\beta$ are forces of inertia that can be derived from the kinetic energy expression (128)

$D_\alpha = - d_{\alpha\beta}\ddot{x}_\beta$ are non-conservative forces that are proportional to the velocities, and they can either diminish the energy of a system (damping) or they can increase this energy (excitation). These forces can be derived from the Rayleigh function

$$D = \frac{1}{2} d_{\alpha\beta}\dot{x}_\alpha\dot{x}_\beta \tag{131}$$

If D is a positive definite quadratic form, then energy is being removed from the system. In that case D is also known as a dissipation function.

$G_\alpha = - g_{\alpha\beta}\ddot{x}_\beta$ are gyroscopic forces. No matter what the motions of the system may be, these forces will not perform any work and they are therefore conservative forces.

$F_\alpha = - f_{\alpha\beta}\ddot{x}_\beta$ are conservative positional forces (constraining forces) that can be derived from the potential

$$U = \frac{1}{2} f_{\alpha\beta}x_\alpha x_\beta \tag{132}$$

If U is a positive definite quadratic form in the neighbourhood of an equilibrium position $x_\alpha = 0$,then the equilibrium position in question is statically stable (Dirichlet's theorem).

$E_\alpha = - e_{\alpha\beta} \ddot{x}_\beta$ are non-conservative positional forces which, just like the D_α forces, can bring about a change in the energy of the system. These forces have also been referred to as circulatory forces.

The forms (127) and (130) of the equations of motion will be used in Section 5.3 for the purposes of examining the transient behaviour and the stability of gyro systems.

5.2.2 Systems with Fast-Running Gyros

Gyroscopic instruments for technological purposes generally contain fast-running gyros, this being due to the fact that it is desired to obtain the angular momentum needed for the proper operation of the instrument with as small a weight as possible. When the gyroscopic forces predominate, it is generally possible to effect some very considerable simplifications in the calculations. Approximations of this kind are often collectively referred to as constituting a "technological gyro theory" or a "technological theory of spinning bodies". It is here proposed to show what assumptions lie at the basis of such approximations and also to discuss the limits of this method.

When the angles φ_κ of the rotation of the various gyros contained in the system about their own principal axes of symmetry (principal moments of inertia C_κ) are used as cylical coordinates, the kinetic energy of the system can be written in the form

$$T = \frac{1}{2} a_{\alpha\beta} \dot{q}_\alpha \dot{q}_\beta + \frac{1}{2} C_\kappa (\dot{\varphi}_\kappa + h_{\kappa\alpha} \dot{q}_\alpha)^2 \tag{133}$$

In this expression the suffices α and β have values that run from 1 to m, while the suffix κ has values running from m + 1 to n. The magnitude $h_{\kappa a}$ is equal to the cosine of the angle between the direction of the axis of symmetry of the κ-th gyro and the direction of the vectors of angular velocity $\dot{q}_a$. The values of α cover all those rotations that are available as carrying rotations for the κ-th gyro. The expression in the rounded brackets of (133) is therefore equal to the component of the absolute rotation of the κ-th gyro in the direction of its own axis of symmetry.

Having chosen the angles φ_κ as cyclical coordinates, it follows that

$$\frac{\partial T}{\partial \dot{\varphi}_\kappa} = C_{(\kappa)} \; (\dot{\varphi}_\kappa + h_{\kappa a} \dot{q}_a) = H_\kappa = \text{const.} \qquad (134)$$

The suffix k in the above has been placed in round brackets in order to indicate that, by way of exception, it is not to be comprised in the summation; indeed, this notation will be used from here onwards. Putting (134) into (133), we now obtain

$$T = \frac{1}{2} a_{\alpha\beta} \dot{q}_\alpha \dot{q}_\beta + \frac{1}{2} C_\kappa \left(\frac{H_\kappa}{C_\kappa}\right)^2 .$$

In accordance with (112) we now obtain the Routhian function as follows

$$R = T - H_\kappa \dot{\varphi}_\kappa = \underbrace{\frac{1}{2} a_{\alpha\beta} \dot{q}_\alpha \dot{q}_\beta}_{R_2} + \underbrace{H_\kappa h_{\kappa a} \dot{q}_a}_{R_1} - \underbrace{\frac{1}{2} \frac{H_\kappa^2}{C_\kappa}}_{R_0} . \qquad (135)$$

Since R_o is constant, the equation of motion (116) now assumes the form

$$(136)\quad \frac{d}{dt}\left(\frac{\partial R_2}{\partial \dot{q}_a}\right) - \frac{\partial R_2}{\partial q_a} = Q_a - \frac{d}{dt}\left(\frac{\partial R_1}{\partial \dot{q}_a}\right) + \frac{\partial R_1}{\partial q_a}\,, \quad (\alpha = 1,\ldots,m).$$

The above equations, which are still general valid, can now be simplified for the case of fast-running gyros by considering that the energy component R_2 can be neglected when compared with R_1. From a physical point of view this means that the kinetic energy produced by the rotations $\dot{q}_a$, that is to say, principally the energy of motion contained in the masses of the gyro housing and the suspensions, is regarded as negligibly small in comparison with the kinetic energy of the gyro rotors. Such an assumption is justified in the case of a large number of technological gyroscopic instruments. Introducing this simplification into the system of equations defined by (136), one obtains the approximate form

$$(137)\qquad \frac{d}{dt}\left(\frac{\partial R_1}{\partial \dot{q}_a}\right) - \frac{\partial R_1}{\partial q_a} = Q_a\,, \quad (\alpha = 1,\ldots, m)$$

In other words, the approximate equations of motion are obtained quite simply by introducing into the Lagrangian equations of motion of the second type not the overall kinetic energy, but merely the R_1 component of the Routhian function, this component depending in a linear manner on the system velocities $\dot{q}_a$, i.e.

$$(138)\qquad R_1 = H_\kappa h_{\kappa a} \dot{q}_a$$

Furthermore, equations (137) can also be brought into the form

$$H_\kappa\left(\frac{\partial h_{\kappa\beta}}{\partial q_\alpha} - \frac{\partial h_{\kappa\alpha}}{\partial q_\beta}\right)\dot{q}_\alpha = g_{\alpha\beta}\,\dot{q}_\alpha = Q_\beta \;, \quad (\beta = 1, \ldots, m) \qquad (139)$$

In this form one can immediately recognize the skew symmetry of the matrix $g_{\alpha\beta}$ of the gyroscopic forces. Equation (139) expresses the fact that the motions are primarily determined by the equilibrium between the external forces Q_α and the gyroscopic forces G_α.

When the approximate equations are being utilized for practical applications, it becomes possible to introduce a further simplification. This consists of basing the calculation of the numerical values of the expressions for the angular momentum H as given by (134) on the assumption that the carrying components can be neglected when compared with the rotation itself, i.e.

$$\dot{\varphi}_\kappa >> h_{\kappa\alpha}\,\dot{q}_\alpha \;.$$

Gyro calculations for technological purposes can therefore be organized into the following steps:

1) Calculate the constant angular momentums given by

$$H_\kappa \approx C_{(\kappa)}\dot{\varphi}_\kappa \quad ;$$

2) Use this to determine the abbreviated Routhian function R_1 in accordance with (138);

3) Formulate the equations of motion in accordance with (137);

4) Solve the equations of motion.

The approximate equations of motion that can be obtained in this manner are of a lower degree than the exact equations. This means that the solutions of the approximate equations cannot be made to fit all arbitrarily chosen initial conditions, and it therefore follows that they cannot be used for calculating all possible conditions of motion of a gyro system. In particular, in view of the fact that these equations neglect the components of the angular momentum due to the additional masses, they will not succeed in giving a picture of the rapid nutational oscillations. On the other hand, they will generally reproduce the slow precessional motions with a very considerable degree of accuracy. For this reason the approximate equations are sometimes called the equations of the precession theory.

5.3. The Transient Behaviour of Gyro Systems

The above-discussed approximate equations of motion will now be used to examine two particular problems that are of great importance for the technology of gyro systems. I am referring, firstly, to the influence of the various types of forces on the stability of the motions and, secondly, to the dependency of the eigenfrequencies on the magnitude of the angular momentum.

5.3.1. Theorems about the Stability of Linear Systems.

In the equations of motion (130) there appear the five types of forces A_a, D_a, G_a, F_a and E_a that have already been appropriately characterized in Section 5.2. Their influence on the stability behaviour of gyro systems has been investigated by numerous authors. The following remarks will endeavour to give a summary picture of the most important results obtained in the course of these investigations. With a view to introducing a certain order into the multiplicity of systems with different combinations of forces, it is proposed to classify the various gyro systems in accordance with the following table.

		$g \equiv o$		$g \neq o$	
		$d \equiv o$	$d \neq o$	$d \equiv o$	$d \neq o$
$f \equiv o$	$e \equiv o$	A	AD	AG 1	ADG 2
	$e \neq o$	AE	ADE	AGE	ADGE 3 (4) 5
$f \neq o$	$e \equiv o$	AF 6	ADF (7)	AGF 8 9 10	ADGF 11 12 13 14 15
	$e \neq o$	AFE 16	ADFE 17	AGFE	ADGFE (18) (19) (20)

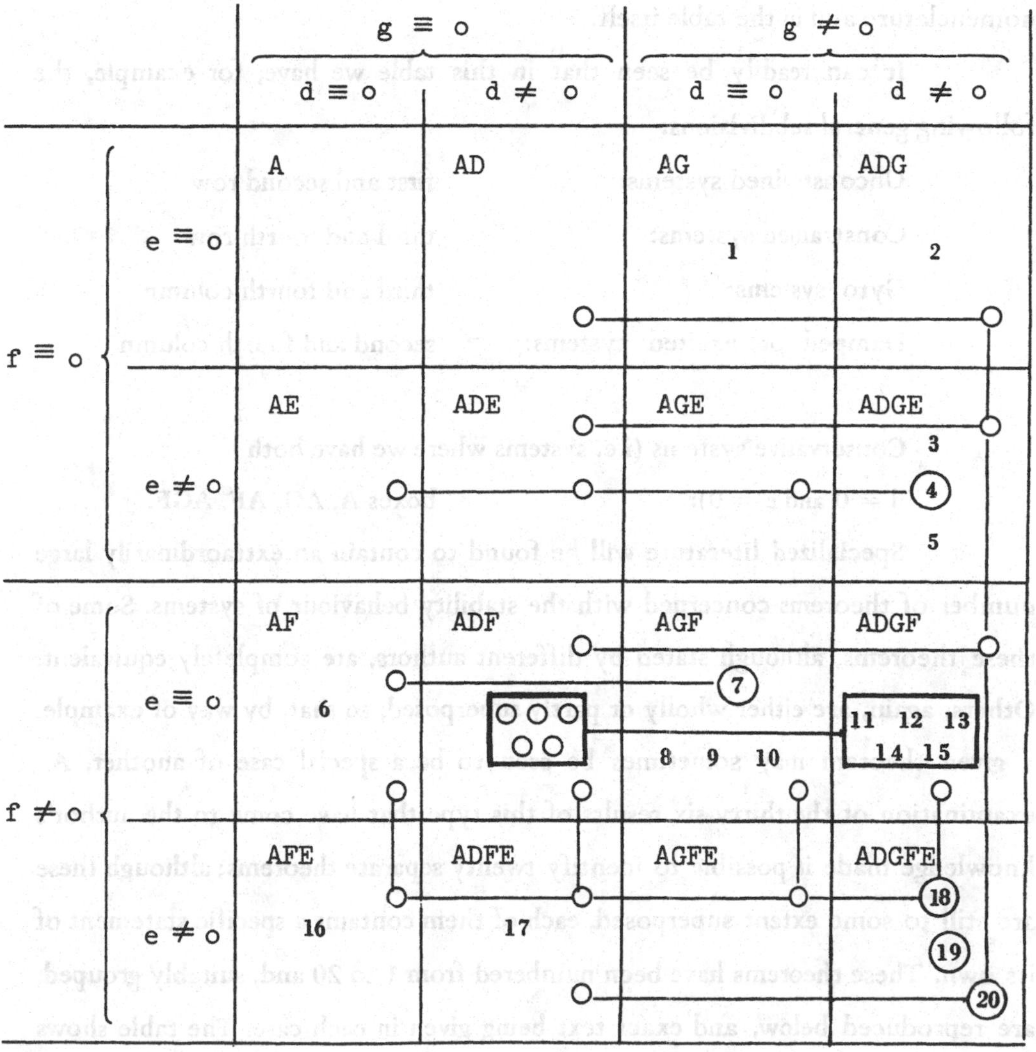

It will be assumed that inertia forces A_a are always present. According to whether the other types of forces are or are not present, the various systems are now univocally characterized by means of a particular box of the table; moreover, each system is assigned a self-explanatory nomenclature describing the forces that are present in it. The use of suffices can be avoided both in the nomenclature and in the table itself.

It can readily be seen that in this table we have, for example, the following general subdivisions:

Unconstrained systems:	first and second row
Constrained systems:	third and fourth row
Gyro systems:	third and fourth column
Damped or excited systems:	second and fourth column

Conservative systems (i.e. systems where we have both $d \equiv 0$ and $e \equiv 0$): boxes A, AG, AF, AGF.

Specialized literature will be found to contain an extraordinarily large number of theorems concerned with the stability behaviour of systems. Some of these theorems, although stated by different authors, are completely equivalent. Others, again, are either wholly or partly superposed, so that, by way of example, a given theorem may sometimes be seen to be a special case of another. An examination of the thirty-six results of this type that have come to the author's knowledge made it possible to identify twenty separate theorems; although these are still to some extent superposed, each of them contains a specific statement of its own. These theorems have been numbered from 1 to 20 and, suitably grouped, are reproduced below, and exact text being given in each case. The table shows the types of system to which the various theorems apply. Some of the theorems mentioned are of a general nature; they are therefore valid for several types of systems and have to be shown in several of the boxes contained in the table. This has been indicated by means of the interconnecting lines, but the actual reference number of the theorem is only shown in the box representing the most general system to which the theorem in question applies.

The proofs of the individual theorems make use of the properties of the quadratic forms T (128), D (131) and U (132), as well as those of the skew symmetrical matrices $g_{\alpha\beta}$ and $e_{\alpha\beta}$. It is not proposed to repeat these proofs here; the reader can readily refer to the appropriate literature, the reference being given in each case. It should be expressly pointed out that these references do not generally refer to the first publication made by the discoverer himself, but rather to summarizing texts that are either more readily accessible or contain more up-to-date formulations of the theorem.

The twenty theorems listed hereinbelow have been reproduced with the text in such a form as to make it possible for the various theorems to be used without having to refer back to the original publications. Although endeavours have made to obtain a certain unity in the formulations, the statements made by the quoted authors have not been changed. By way of supplement, additional explanations are given in some of the cases. Moreover, with a view to making it easier for the reader to visualize matters, a concrete example is given for each of the systems mentioned in the list.

AG Systems (Example: A spinning rigid body in a no-gravity state or a non--propelled space ship, always provided that the gravitational gradient can be neglected).

1) The equilibrium position of a conservative system that is subjected only to

the actions of gyroscopic forces G_a is stable when and only when the condition det $(g_{\alpha\beta}) \neq 0$ is satisfied [13; p.126].

It immediately follows from this theorem that a system that satisfies the conditions of Theorem 1 will always be unstable when the number m of the non-cyclical positional coordinates is odd. This is due to the fact that the determinant always disappears when a matrix with a skew symmetry is of an odd order.

ADG Systems (Example: A spinning body in a no-gravity state subjected to the actions of damping forces that are proportional to the velocities).

2) The equilibrium position of a system that is only subjected to the action of the gyroscopic and dissipative forces G_a and D_a is always stable when the Rayleigh function D (131) is positively definite [13; p.163].

A positive definite Rayleigh function D means that the motions are damped in <u>all</u> the degrees of freedom of the system. In that case the forces D_a act in dissipative rather than an exciting manner.

ADGE Systems (Example; A spinning space ship in a no-gravity state whose position is controlled by means of E-type forces; with the help of such forces it is possible to make the axis of the angular momentum point in any desired direction).

3) When m is even, i.e. when the number of non-cyclical positional coordinates is even, the motions of a system where $F_a \equiv 0$ can be made asymptotically stable if, in addition to the dissipative forces D_a, gyroscopic forces G_a are also brought into play [13; p.214].

4) When the values of m are odd, the motions of a system where $F_a \equiv 0$ cannot be made asymptotically stable, and this quite irrespective of the dissipating,

exciting or gyroscopic forces D_a and G_a that may be acting [13; p.214].

5) The unstable equilibrium position of a non-conservative system where $F_a \equiv 0$ can only be stabilized if both dissipative and gyroscopic forces D_a and G_a are brought into play [14; p.33].

AF Systems (Example: an undamped oscillator consisting of a spring and a mass).

6) The characteristic equation of a conservative system in which $G_a \equiv 0$ and potential energy U (132) is non-definite possesses at least one root with a positive real part.

The characteristic equation of an AF system has the form $\det(a_{\alpha\beta}\lambda^2 + f_{\alpha\beta}) = 0$. Since one can always carry out a coordinate transformation in such a way as to bring both the matrices $a_{\alpha\beta}$ and $f_{\alpha\beta}$ into a diagonal form it follows that in the event of a non-definite potential energy at least one of the diagonal elements of the transformed constraint matrix $f_{\alpha\beta}$ must be negative. This immediately implies the existence of a root of the characteristic equation where $\mathrm{Re}(\lambda) > 0$.

AGF Systems (Examples: A gyro pendulum, a monorail without any damping devices, or a frictionless toy top).

7) The characteristic equation of a conservative system with a stable equilibrium position possesses only roots in which the real parts disappear [15; p.49].

A stable equilibrium position is characterized by a positively definite potential energy U. The theorem reproduced above, which was already formulated by Lagrange, states that a conservative system will always oscillate in an undamped manner when a stable equilibrium is disturbed.

8) When the equilibrium position of a conservative system is characterized by the presence of an odd number of degrees of freedom, then this equilibrium position cannot be stabilized by the addition of gyroscopic forces G_a [13; p. 176].

The number of the unstable degrees of freedom is equal to the number of roots λ with $Re(\lambda) > 0$.

9) The unstable equilibrium position of a conservative system can be stabilized by gyroscopic forces G_a when 1) $\det(g_{\alpha\beta}) \neq 0$, 2) the potential energy U is definite, 3) the characteristic equation does not possess any multiple roots, and 4) the angular momentum H is sufficiently great [13; p. 179].

Theorems 8 and 9 contain a classical result that is due to Thomson and Tait [16]. However, the practical importance of this statement is not very great because it only applies in the case of conservative systems. In real systems there always exist a loss of energy; as a result of this there can be fundamental changes in the stability behaviour, as will be shown by Theorem 11 below.

10) A conservative system with a stable equilibrium position cannot be made unstable by the addition of gyroscopic forces G_a [15; p. 49].

ADGF Systems (Example: A damped gyro pendulum).

11) When the Rayleigh function D is positive definite and the constraint matrix $f_{\alpha\beta}$ does not have any disappearing eigenvalues, then the stability is independent of the damping and gyroscopic forces D_a and G_a [17; p. 47].

This theorem is extremely important. It was already suspected by Thomson and Tait [16], and subsequently formulated and proved by Chetajew [18]. In the last resort it states that the stability of a complete ADGF system characterized by the equation of motion

$$a_{\alpha\beta}\ddot{x}_\beta + d_{\alpha\beta}\dot{x}_\beta + g_{\alpha\beta}\dot{x}_\beta + f_{\alpha\beta}x_\beta = 0 \quad (\alpha,\beta = 1,2,\ldots, m) \qquad (140)$$

corresponds to the stability of an AF system with the abbreviated equation of motion

$$a_{\alpha\beta}\ddot{x}_\beta + f_{\alpha\beta}x_\beta = 0 \quad (\alpha,\beta = 1,2,\ldots, m) \qquad (141)$$

It therefore follows that in the last analysis the stability of an ADGF system is determined exclusively by the eigenvalues of the constraint matrix $f_{\alpha\beta}$, these eigenvalues being solutions of the equation $\det(f_{\alpha\beta} - \lambda\delta_{\alpha\beta}) = 0$. Theorem 11 demonstrates that a stable system can be made unstable by the addition of damping forces D_a. In fact, it can be proved that the upright Lagrangian gyro is unstable when one considers the damping influences and applies the previous explained concept of stability, this being also true in the case of the toy top. Nor is this contradicted by the fact that damping influences cause the toy top to tend towards an upright position and thus lead to a dying out of precessional motions. In any case, when $t \to \infty$, the angular momentum of the toy top becomes diminished to such an extent that the top collapses to the ground.

12) When the Rayleigh function is positively definite, then the number of roots of the characteristic equation with a positive real part is equal to the number of the negative eigenvalues of the constraint matrix $f_{\alpha\beta}$ [17; p. 47].

13) Statically stable equilibrium positions remain stable, and this even when any arbitrarily chosen gyroscopic and damping forces G_a and D_a with a positively semi-definite Rayleigh function D are added [13; p. 202].

When the Rayleigh function D is positive semi-definite, the system can perform undamped oscillations in one or more degrees of freedom, while all other motions are subjected to damping.

14) When the Rayleigh function D is positive semi-definite, the motions of a system are unstable, no matter what gyroscopic forces G_a may come into play, always provided that all the eigenvalues of the constraint matrix $f_{\alpha\beta}$ are negative [17; p. 48].

15) A statically stable equilibrium position becomes asymptotically stable when damping forces D_a with a positively definite Rayleigh function D and arbitrarily chosen gyroscopic forces G_a are added [13; p. 203].

AFE Systems (Example: A double pendulum in which the lower pendulum is subject to a force acting in a direction that remains fixed with respect to the body, i.e. rotates with it).

16) The equilibrium position of a conservative system in which $G_a \equiv 0$ can be stabilized or made unstable by the addition of non-conservative positional forces E_a [19; p. 108].

ADFE Systems (Example: A double pendulum as in the case of the AFE system, but subject either to damping or to excitation).

17) When the equilibrium position of a system in which $G_a \equiv 0$ is statically unstable, then such an equilibrium position cannot be stabilized by the addition of non-conservative positional forces E_a [14; p. 33].

ADGFE Systems (Examples: A gyro pendulum with a Sperry-type control device and damping, a monorail with a stabilization device, or a rotating satellite with attitude control in an orbit close to the earth).

18) The equilibrium positions of a system with a negative definite potential energy U cannot be stabilized by the addition of arbitrarily chosen forces D_a, G_a or E_a when the number of non-cyclical positional coordinates is odd [13; p.215].

19) When the number of non-cyclical positional coordinates is even and damping forces D_a with a positively definite Rayleigh function D are acting, the equilibrium positions of a system with a negative definite potential energy U can only be stabilized if both gyroscopic forces G_a and non-conservative positional forces E_a are brought into play [13; p.215].

20) The motions of a system are unstable when the trace of the matrix $d_{\alpha\beta}$ is negative, and this no matter what forces G_a, F_a and E_a may be acting [13; p.213].

The condition $tr(d_{\alpha\beta}) = d_{\alpha\alpha} > 0$ has the physical meaning that the damping components in the Rayleigh function D predominate when compared with the excitation components. This can readily be seen as follows: if a coordinate transformation is used to bring the matrix $d_{\alpha\beta}$ into a diagonal form, the positive elements of the transformed matrix refer to damped forms of oscillations, while the negative elements refer to excited forms. When the sum of the positive elements is greater than the sum of the negative ones, then damping predominates. However, since the sum of the diagonal elements of a matrix (i.e. its trace) remains invariant with respect to coordinate transformations, the predominance or non-predominance of the damping components can be ascertained by forming the trace of the original matrix $d_{\alpha\beta}$.

5.3.2. The Dependency of the Eigenfrequencies on the Angular Momentum.

In this section we shall limit ourselves to considering conservative systems, because such systems permit one to gain a readier insight into the various relationships. In place of (130) we shall therefore presuppose a system of equations of the form

$$a_{\alpha\beta}\ddot{x}_\beta + g_{\alpha\beta}\dot{x}_\beta + f_{\alpha\beta}x_\beta = 0\,, \quad (\alpha = 1, \ldots, m) \tag{142}$$

The roots $\pm\lambda_\delta(\delta=1,\ldots,m)$ of the characteristic equation of this system, i.e.

$$\det\,(a_{\alpha\beta}\lambda^2 + g_{\alpha\beta}\lambda + f_{\alpha\beta}) = 0 \tag{143}$$

are purely imaginary (see Theorem 7 in Section 5.3.1.) and consist of pairs that differ only as far as the sign is concerned, i.e.

$$\lambda_\delta = \pm\, i\omega_\delta \qquad (\delta = 1, \ldots, m). \tag{144}$$

This means that the motion consists of a superposition of undamped oscillations. Without solving the characteristic equation of the system and thus explicity calculating the circular frequencies of these oscillations, we can nevertheless arrive at a number of very general statements. This can be done as follows:

Assuming that the characteristic equation does not possess any multiple roots, the complete solution must be made up of the partial solutions

$$x_{\beta\delta} = A_{\beta(\delta)}\,e^{\lambda_\delta t} \tag{145}$$

If these are substituted in the original equations (142), they yield the system of equations

$$(a_{\beta\gamma}\lambda_\delta^2 + g_{\beta\gamma}\lambda_\delta + f_{\beta\gamma})\,A_{\beta(\delta)} = 0\,, \quad (\gamma = 1, \ldots, m) \tag{146}$$

The amplitude factors $A_{\beta\delta}$ in the above can be complex. Since the coefficients of the system of equations are real, the amplitude factors pertaining to the two eigenvalues (144) must be conjugate with respect to each other. We therefore have

$$A_{\beta\delta} = B_{\beta\delta} + i\, C_{\beta\delta}\ ; \qquad A_{\beta(m+\delta)} = B_{\beta\delta} - i\, C_{\beta\delta} = \bar{A}_{\beta\delta}\ . \tag{147}$$

We now multiply the equations of the system (146) by $\bar{A}_{\gamma\delta}$ and then sum over γ, obtaining

$$(a_{\beta\gamma}\lambda_\delta^2 + g_{\beta\gamma}\lambda_\delta + f_{\beta\gamma})\, A_{\beta(\delta)}\, \bar{A}_{\gamma(\delta)} = 0\ . \tag{148}$$

Taking into consideration (147) and the symmetry of the matrices $a_{\beta\gamma}$ and $f_{\beta\gamma}$, as well as the skew symmetry of the matrix $g_{\beta\gamma}$, we now obtain the following for the expressions that occur in (148)

$$\left.\begin{aligned} a_{\beta\gamma} A_{\beta(\delta)} \bar{A}_{\gamma(\delta)} &= a_{\beta\gamma}\left[B_{\beta(\delta)}B_{\gamma(\delta)} + C_{\beta(\delta)}C_{\gamma(\delta)}\right] = a_\delta > 0\ ,\\ g_{\beta\gamma} A_{\beta(\delta)} \bar{A}_{\gamma(\delta)} &= i g_{\beta\gamma}\left[C_{\beta(\delta)}B_{\gamma(\delta)} - C_{\gamma(\delta)}B_{\beta(\delta)}\right] = i g_\delta\ ,\\ f_{\beta\gamma} A_{\beta(\delta)} \bar{A}_{\gamma(\delta)} &= f_{\beta\gamma}\left[B_{\beta(\delta)}B_{\gamma(\delta)} + C_{\beta(\delta)}C_{\gamma(\delta)}\right] = f_\delta\ . \end{aligned}\right\} \tag{149}$$

In the above a_δ, g_δ, f_δ represent real magnitudes; the a_δ terms, being the sum of two positive definite quadratic forms, are always positive; the f_δ terms, likewise, are the sums of quadratic forms, but they are only positive in the case of statically stable equilibrium positions; the g_δ terms, on the other hand, are bi-linear forms and can thus be both positive and negative. The equation (148) now assumes the form

$$a_{(\delta)}\lambda_\delta^2 + i g_{(\delta)}\lambda_\delta + f_\delta = 0\ , \tag{150}$$

and its solutions are represented by

(151) $$\lambda_\delta = - i \frac{g_{(\delta)}}{2a_{(\delta)}} \left[1 \pm \sqrt{1 + \frac{4a_{(\delta)} f_{(\delta)}}{g^2_{(\delta)}}} \right]$$

From this it can readily be seen that the condition

(152) $$g^2_\delta + 4 \ a_{(\delta)} f_\delta > 0 \qquad (\delta = 1, \ldots, m)$$

must necessarily be satisfied if the solutions λ_δ are to be purely imaginary.

From the above, inter alia, there immediately follows the result that has already been stated in Theorems 8 and 9 of Section 5.3.1, i.e. that there is always stability in the presence of statically stable constraints ($f_\delta > 0$), while in the case of statically unstable constraints ($f_\delta < 0$) stable motions can be obtained only when the angular momentum is sufficiently great.

We now propose to extend the results we have just obtained so as to cover the case of very great angular momentum ($H \to \infty$),. For this purpose we introduce an angular momentum parameter H as defined by

$$H_\kappa = H \ k_\kappa$$

From this it immediately follows that we must also have

(153) $$g_{\alpha\beta} = H \ g^\star_{\alpha\beta} \quad \text{and} \quad g_\delta = H \ g^\star_\delta$$

Since the angular momentum appears in linear form in the elements of $g_{\alpha\beta}$, the magnitudes $g^\star_{\alpha\beta}$ and $g^\star_\delta$ in the above are independent of H.

If we now substitute (153) in (151) and bear in mind the fact that $H \to \infty$, the root occurring in (151) can be developed into a series. Neglecting in each case all the terms of the developed series other than the one that contains the highest power of H, we obtain the two approximate values

$$\left.\begin{aligned} \lambda_{\delta 1} &\simeq i\,\frac{f_{(\delta)}}{H\,g_\delta^\star}\,, \\ \lambda_{\delta 2} &\simeq -i\,\frac{H\,g_\delta^\star}{a_{(\delta)}}\,, \end{aligned}\right| \qquad (\delta = 1, \ldots, m) \qquad (154)$$

As $H \to \infty$, the roots $\lambda_{\delta 1}$, being proportional to 1/H, tend towards zero; they represent the slow precession oscillations of a gyro system. The roots $\lambda_{\delta 2}$, on the other hand, increase in magnitude in direct proportion to H. They correspond to the rapid nutational oscillations. We can therefore formulate the following general theorem:

> The frequencies occurring in a stable, conservative and gyroscopic system in which $\det(g_{\beta\gamma}) \neq 0$ are always divided into two groups: when the angular momentum is very great, the frequencies of the one group are inversely proportional to the angular momentum, while those of the other group change in direct proportion to it.

It should be quite obvious that such a division of the frequencies into two groups is only possible when m is even.

A more detailed examination of the characteristic equation (143) shows that not only m must be even, but that the additional condition $\det(g_{\alpha\beta}) \neq 0$ must also be complied with. The characteristic equation can be written as a polynomial, i.e.

$$\Delta(\lambda) = b_{2m}\lambda^{2m} + b_{2m-2}\lambda^{2m-2} + \ldots + b_2\lambda^2 + b_0 = 0 \qquad (155)$$

In this polynomial the coefficients

$$b_{2m} = \det(a_{\alpha\beta}) \quad \text{and} \quad b_o = \det(f_{\alpha\beta}) \qquad (156)$$

are independent of H. If one ignores some special cases, the powers of H in the coefficients b will increase as the distance of the coefficient from the terminal members of the polynomial increases. The highest power of H is to be found in the coefficient b_m because this coefficient contains the term $H^m \det(g^*_{\alpha\beta})$. In the limiting case, i.e. when $H \to \infty$, the dominant components everywhere will be those that contain H as a factor. Since adjacent terms of the polynomial (155) always contain different powers of H, it is possible to obtain approximations for the roots by splitting the equation (155) into partial equations that consist in each case of two adjacent terms. On the basis of what has been said before, the first m roots will now be proportional to H, while the remaining m roots will be proportional to 1/H. However, this is no longer true when the matrix $g_{\alpha\beta}$ is singular. For example, if $\det(g_{\alpha\beta}) = 0$ and the first subdeterminants do not disappear, the coefficients b_{m+2}, b_m, b_{m-2} will all contain the same order of H. It follows from this that, when $H \to \infty$, two of the roots of equation (155) will tend towards certain limiting values that are independent of H. These oscillations, which cannot be included either among the nutations or the precessions, will be described quite generally as "pendulum oscillations".

In systems in which m is an odd number there will exist at least one such pendulum oscillation. The greater the defect of the matrix $g_{\alpha\beta}$, the greater will be the number of pendulum oscillations that make their appearance.

Moreover, in the case of a non-singular matrix $g_{\alpha\beta}$, approximate solutions can also be obtained directly from the equations of motion themselves. Since $g_{\beta\gamma} = H g^{\star}_{\beta\gamma}$, the middle term in equation (142) will always become dominant when the angular momentum is very great. When the motions are slow, on the other hand, the forces of inertia involved in them will surely be small. Such motions can therefore be calculated in an approximate manner from the condition of equilibrium between the gyroscopic forces and the constraint forces. These considerations lead to the approximate equations

$$g_{\alpha\beta}\dot{x}_{\beta} + f_{\alpha\beta}x_{\beta} \simeq 0 , \qquad (\alpha = 1, \ldots, m). \qquad (157)$$

It can be seen that we have thus arrived once again at the approximate equations (139), equations that were previously deduced in a completely different manner. Their solution, leaving aside some exceptional cases, yields the roots $\lambda_{\delta 1}$ of (154), i.e. the roots that correspond to precessions. For this reason the equation (157) are also referred to as the precession equations of the theory of spinning bodies.

On the other hand, when it is desired to examine rapid motions (nutations), the forces of inertia will undoubtedly be substantially greater than the constraint forces. Basing oneself on (142) and considering that there must now be an equilibrium between the forces of inertia and the constraint forces, one thus obtains the approximate equations

$$a_{\alpha\beta}\ddot{x}_{\beta} + g_{\alpha\beta}\dot{x}_{\beta} \simeq 0, \qquad (\alpha = 1, \ldots, m) \qquad (158)$$

Their solution, again leaving aside some exceptional cases, yields the roots $\lambda_{\delta 2}$ of (154), i.e. the roots corresponding to nutations.

6. Friction Effects

6.1. Effect produced by Friction in the Case of a Cardanic Suspended Gyro

According to the type of friction and the place where it comes into play, a cardanic suspended gyro will become subject to various types of phenomena that are of importance as regards the technical applications of such gyros.

We shall begin by listing a number of results that can be proved both theoretically and experimentally for the case where friction comes into play in the gimbal bearings of cardanic suspended gyro pendulum, although the proofs will not be given here. The following statements can be made both in the case of viscous friction, i.e. when the damping is proportional to the angular velocities, and in the case of constant Coulomb friction:

1) The precession motions of a hanging gyro, i.e. one that has statically stable constraints, become damped.
2) The precession motions of an upright (or statically unstable) gyro will have increasing amplitudes.
3) The precession motions of gyro pendulum with mixed constraints (i.e. statically stable constraints with respect to one gimbal axis, and unstable with respect to the other gimbal axis) are asymptotically unstable .
4) The nutation motions are damped in every case. If the damping is of the viscous kind, the oscillations will completely cease after an appropriately long time. In the case of Coulomb friction, on the other hand, there may be an undamped residual oscillation if additional torques are acting about one of the gimbal axes. By way of typical example, we now propose to make a more detailed examination of this particular effect.

Before doing so, however, it should be pointed out that the four effects we have listed do not by any means exhaust the phenomena that may occur with cardanic suspended gyros. For example, in the presence of Coulomb friction it is also possible to have a kind of stick-slip motion, that is to say, a temporary blockage of motion about one or both of the gimbal axes. There is a particularly great multiplicity of types of motion when the gyros are located in rotating reference systems, on the earth for example. In this connection the reader is referred to the more advanced works of Grammel and Ziegler [20] and Butenin [21].

6.1.1. Nutations in the Presence of Coulomb Friction

It is now proposed to investigate the nutational oscillations of a cardanic suspended gyro whose gimbal bearings transmit friction torques of the Coulomb type and whose outer gimbal axis is subject to the additional action of a constant torque M_o. In this case the equations of motion, linearized for small motions, assume the following form

$$\begin{aligned} A\ddot{\alpha} + H\dot{\beta} &= M_1 = - r_1 \operatorname{sgn} \dot{\alpha} + M_o \, , \\ B\ddot{\beta} - H\dot{\alpha} &= M_2 = - r_2 \operatorname{sgn} \dot{\beta} \, . \end{aligned} \qquad (159)$$

Given the discontinuity of the sign function, it is not possible to obtain a unique analytical solution. Nevertheless, a complete picture of the motions can be obtained when one examines the state curves in a particular velocity plane. This can best be done by plotting $\dot{\beta}$ on the abscissa of the standardized angular velocities given by $\dot{\alpha}^* = \sqrt{A/B}\,\dot{\alpha}$, because in this case the curves assume a particularly simple form. From (159) one first obtains

$$\sqrt{AB}\,\ddot{\alpha}^* = - H\dot{\beta} - r_1 \operatorname{sgn} \dot{\alpha}^* + M_o \, , \qquad (160/1)$$

$$\text{(160/2)} \qquad \sqrt{AB}\,\ddot{\beta} = H\ddot{\alpha}^* - \sqrt{\frac{A}{B}}\, r_2 \operatorname{sgn} \dot{\beta} \quad .$$

and from this it follows that

$$\text{(161)} \qquad \frac{\ddot{\beta}}{\ddot{\alpha}^*} = \frac{d\dot{\beta}}{d\dot{\alpha}^*} = -\frac{H\dot{\alpha}^* - \sqrt{\frac{A}{B}}\, r_2 \operatorname{sgn} \dot{\beta}}{H\dot{\beta} + r_1 \operatorname{sgn} \dot{\alpha}^* - M_0} \quad .$$

The fraction that forms the right-hand side of this equation is a function of $\dot{\alpha}^*$ and $\dot{\beta}$. Equation (161) therefore assigns a particular direction to each point of the $\dot{\alpha}^*,\dot{\beta}$-plane, and this makes it possible to construct or calculate the state curves. Singular points of the state plane appear when both the numerator and the denominator disappear. This condition is complied with when

$$\text{(162)} \qquad \dot{\alpha}^*_o = \frac{r_1}{H}\sqrt{\frac{A}{B}} \operatorname{sgn} \dot{\beta} \; ; \; \dot{\beta}_o = -\frac{r_1}{H} \operatorname{sgn} \dot{\alpha}^* + \frac{M_o}{H} \quad .$$

To each quadrant there now corresponds a unique sign combination of sgn $\dot{\alpha}^*$ and sgn $\dot{\beta}$ and consequently also an equilibrium position that is valid within this quadrant. This is illustrated in Fig. 14, where each quadrant, identified by means of a Roman numeral, has been assigned the corresponding equilibrium position, these latter being designated by means of the appropriate Arab numerals.

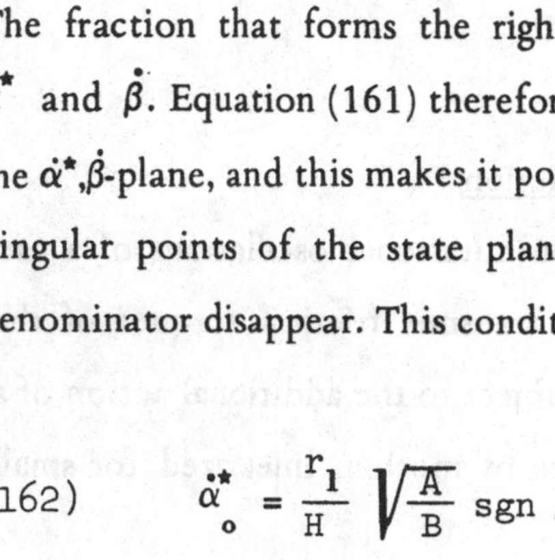

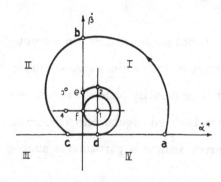

Fig. 14

In order to obtain the state curves in the $\dot{\alpha}^*$, $\dot{\beta}$-plane, one now performs the transformation

$$u = \ddot{\alpha}^* - \ddot{\alpha}^*_0 \quad ; \quad v = \dot{\beta} - \dot{\beta}_0 \qquad (162a)$$

In this way equation (161) becomes transformed into

$$\frac{dv}{du} = -\frac{u}{v} \quad \text{or} \quad udu + vdv = 0 .$$

Integration of the above yields the equation of a circle located in the u,v-plane, i.e.

$$u^2 + v^2 = \text{const} = r^2 \qquad (163)$$

In each quadrant one therefore obtains as the state curve an arc of a circle that has its centre at the singular point that characterizes the quadrant in question. The complete state curve can then be constructed by joining the individual circular arcs. It can readily be ascertained from (159) that the representing points run through the circular arcs in a counterclockwise direction. In this way, for example, one obtains the kind of state curve of which Fig. 14 shows a typical example. Starting from Point a , the curve proceeds via b and eventually reaches c. There it continues as a straight line until d is reached, because the state point is now characterized by a scattering motion and slides along the quadrant boundary. Between d and e the curve again assumes the form of a circular arc, between e and f we have another scatter line, and lastly, from f onwards, a circle round equilibrium position 1 with a radius such that the curve never goes beyond the bounds of Quadrant 1. The residual circle can be interpreted as the limit cycle of the non-linear oscillations of the system. This circle is described by

$$\dot{\alpha}^* = \dot{\alpha}_o^*(1 - \cos \omega^N t) \quad , \tag{164}$$
$$\dot{\beta} = \dot{\beta}_o - \dot{\alpha}_o^* \sin \omega^N t)$$

where $\omega^N = H/\sqrt{AB}$ represents the circular frequency of the nutational oscillations of the gyro. Integration of equations (164) then leads to

$$\alpha = \alpha_o + \dot{\alpha}_o^* \left(t - \frac{1}{\omega^N} \sin \omega^N t\right) \quad , \tag{165}$$
$$\beta = \beta_o + \dot{\beta}_o t + \frac{\dot{\alpha}_o^*}{\omega^N} \cos \omega^N t \quad .$$

This shows that the gyro drifts not only about the outer gimbal axis, i.e. the axis about which the additional torque M_o is acting, but rather about both gimbal axes. Undamped nutational oscillations are superposed on this drift motion. The radius of the residual circle diminishes as M_o decreases, and disappears completely when $M_o \leqslant r_1$. When this condition applies, residual nutations are no longer possible.

6.1.2. Gyro Collapse

An effect that is of interest in gyro technology, the so-called gyro collapse, can occur in the case of cardanic suspended gyros when the rotor is decelerating and the inner gimbal has a position where it comes to rest against a stop. When the rotor begins to decelerate, one at first observes an increase of the inclination angle β of the inner gimbal. If the inner gimbal rotates as far as to come against the stop, the outer gimbal will begin to rotate, in most cases rather suddenly, and with increasing speed. Viceversa, when the rotor is accelerating from its rest position, one can note that the inner gimbal has a tendency to assume a normal position with respect to the outer gimbal (i.e. $\beta \to 0$).

For the purpose of calculating these effects, let us consider a 1,2,3-system that remains rigidly fixed with respect to the outer gimbal. If we now assume that the rotor is spinning so fast that the components of the angular momentum of the gimbals can be neglected when compared with the angular momentum of the rotor, then the vector of the angular momentum can be considered to have the components

$$H_i \simeq (H \sin\beta \ , \ 0 \ , \ H \cos\beta) . \tag{166}$$

The equation of motion in the reference system that is rotating at an angular velocity of $\Omega_i = (\dot{\alpha}, 0, 0)$ can then be obtained as

$$\frac{dH_i}{dt} + \epsilon_{ijk} \Omega_j H_k = M_i \ , \tag{167}$$

and, when this is calculated for the first two components, one obtains the two equations

$$\left. \begin{aligned} \dot{H} \sin\beta + H \cos\beta \, \dot{\beta} &= M_1 \ , \\ - H \cos\beta \, \dot{\alpha} &= M_2 \ . \end{aligned} \right| \tag{168}$$

If the system is an astatic one and the gimbal bearings do not have friction, we have $M_1 = M_2 = 0$. It then follows from (168) that

$$\frac{d}{dt} (H \sin\beta) = 0 \ \text{and} \ \dot{\alpha} = 0 .$$

Hence we have

(169) $$H \sin\beta = H_0 \sin\beta_0 \;; \qquad \alpha = \alpha_0 .$$

Since $H = H_3' = C^R\omega_3'$, the motion of the inner gimbal is described by

(170) $$\sin\beta = \frac{H_0 \sin\beta_0}{C^R\,\omega_3'(t)}$$

This confirms the effects that were described at the beginning: in the domain in which we are interested, when $\dot{\beta}_0 > 0$ and the rotor is accelerating from rest, we have $\dot{\omega}_3' > 0$ and consequently $\dot{\beta} < 0$, i.e. the inner gimbal drifts towards its normal position; on the other hand, when the rotor is decelerating, we have $\dot{\omega}_3' < 0$ and consequently $\dot{\beta} > 0$, i.e. the inner gimbal drifts with increasing inclination.

This also makes it possible to explain the "gyro collapse": when the inner gimbal has drifted to such an extent as to come to rest against the stop, then a torque M_2 depending on the pressure exerted by the stop has to be introduced into the second equation of (168). In this way one finds that the angular velocity of the outer frame is given by

(171) $$\dot{\alpha} = -\frac{M_2}{C^R\omega_3'(t)\cos\beta_A}$$

where β_A represents the angular position of the inner gimbal stop. Since $\dot{\alpha}$ increases even further as $\omega_3'(t)$ decreases, it was found necessary to take special measures in some gyroscopic instruments in order to prevent an excessively rapid rotation of the outer gimbal.

The results described above do not undergo any qualitative changes when additional friction torques are acting in the gimbal bearings.

6.2. Friction Effects in Gyros rotating on a Horizontal plane

A rigid body rotating on a horizontal plane, the toy top for example, has a total of five degrees of freedom. For the purpose of calculating its motions we can make use of the theorem of angular momentum, the theorem of momentum, and also the conditions of constraint that can be formulated for the motion of the body relative to the horizontal plane. The body can become displaced with respect to the base and at the same time it can rotate about an axis that is vertical to the base. It therefore follows that both sliding friction and drill friction will come into play. It can further be shown that the immediate influence of drill friction is rather small. However, the vertical rotation leads to a rather remarkable modification of the laws governing the sliding friction. While the sliding friction by itself can be calculated on the basis of the friction laws formulated by Coulomb, the superposition of sliding motions and drilling motions produces a kind of averaging. In the last resort this leads to a state of affairs where the force of sliding friction is approximately proportional to the sliding speed (see, for example, Contensou [22]). This effect, moreover, can be quite readily observed in floor polishing machines that work with rotating brushes. When the brushes are at rest, a considerable force has to be exerted in order to displace the machine (Coulomb's Law). When the brushes are rotating, on the other hand, it becomes possible to effect a slow displacement of the machine almost without exerting any force at all: the sliding friction has been reduced as a result of the superposition of the rotation.

6.2.1. The Equations for the Toy Top

In this section, following the model provided by Contensou and making use of a friction law that expresses the friction force as proportional to the speed (and also covers the two limiting cases that were mentioned at the beginning), we now propose to develop a linear theory for the stability of the rotations of a symmetrical body on a horizontal plane. Let us assume that we have a symmetrical body where $A = B$, and that its centre of gravity S lies on the axis of symmetry 3' (Fig. 15). Let the body be in contact with the plane at the point P, and assume that its surface in the neighbourhood of P approximates to a sphere with radius r. The centre of this sphere, once again, will be taken to lie on the axis of symmetry 3'.

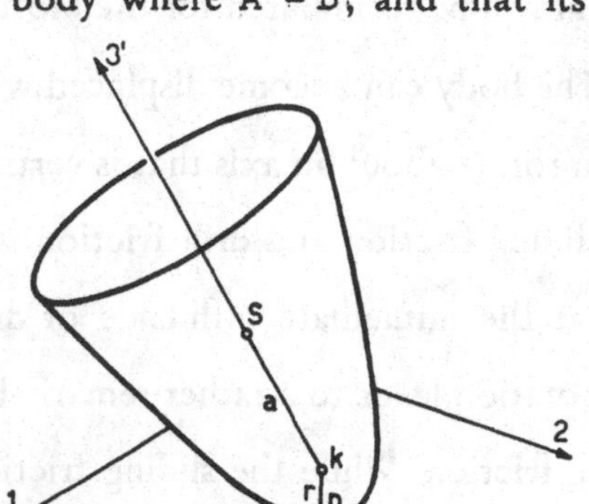

Fig. 15

We shall now choose the horizontal plane as the 1,2-plane of reference system that remains fixed with respect to space. For the purpose of describing the motions of the body we shall make use of the cardanic angles α, β, γ (see Section 2.3.) and the position vector $\overset{S}{x}_i$ of the centre of gravity S. We shall assume both the angles α and β and the changes in $\overset{S}{x}_i$ to be small for the purpose of a first order approximation; this will enable us to perform a linearization of these magnitudes in the sense of the theory of small oscillations.

When the drill friction is neglected, the body can perform a stationary motion that is characterized by $\alpha = \beta = 0, \dot{\gamma} = \omega_0, \overset{S}{x}_i = \overset{S}{x}_{i0}$: in this case the body rotates about the axis of symmetry 3' and this latter is in a vertical position ("sleeping top"). For the immediately adjacent motions one then obtains the linearized system

$$\left.\begin{aligned} \frac{d}{dt}\,(A\dot{\alpha} + C\omega_o\,\beta) &= M_\alpha = M_1\ , \\ A\ddot{\beta} - C\omega_o\dot{\alpha} &= M_\beta \ = M_2\ , \\ \frac{d}{dt}\,(C\dot{\gamma}) &= M_\gamma \ = M_3\ . \end{aligned}\right| \qquad (172)$$

These equations follow from the theorem of angular momentum as formulated for the centre of gravity S. Using the notation shown in Fig. 16, the torques in these equations must therefore be replaced by the expression

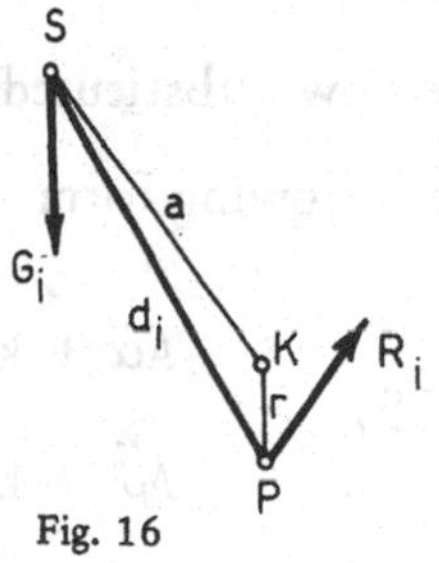

Fig. 16

$$M_i = \epsilon_{ijk}\, d_j\, R_k\ , \qquad (173)$$

where

$$d_j = \begin{bmatrix} -\,a\beta \\ +\,a\alpha \\ -\,h \end{bmatrix} \quad R_\kappa = \begin{bmatrix} R_1 \\ R_2 \\ R_3 \end{bmatrix} = \begin{bmatrix} -\,k\,v_1^{PK} \\ -\,k\,v_2^{PK} \\ G \end{bmatrix}$$

In these expressions $h = a+r$ represents the maximum height of S above P, k is the factor of proportionality of the friction force, and v_i^{PK} is the speed of that point of the body which at that particular moment represents the point of contact P. This latter speed consists of components due to the displacement of the centre of gravity, the rotation about the axes 1 and 2 and the rotation of the point itself about the axis of symmetry, i.e.

(174)
$$v_1^{PK} = v_1^{S} - h\dot{\beta} + r\omega_o\alpha ,$$
$$v_2^{PK} = v_2^{S} + h\dot{\alpha} + r\omega_o\beta .$$

We now put $\dot{\gamma} = \omega_o$ = const. This follows from the first of equations (172) when one considers the fact that in accordance with (173), $M_3 \approx 0$. If (173) and (174) are now substituted in (172), one obtains the first two coordinate equations in the following form

(175)
$$A\ddot{\alpha} + kh^2\dot{\alpha} - aG\alpha + C\omega_o\dot{\beta} + khr\omega_o\beta + khv_2^S = 0$$
$$A\ddot{\beta} + kh^2\dot{\beta} - aG\beta - C\omega_o\dot{\alpha} - khr\omega_o\alpha - khv_1^S = 0$$

Furthermore, from the theorem of momentum one obtains the first two coordinate equations as

(176)
$$m\dot{v}_1^S = R_1 = -k(v_1^S - h\dot{\beta} + r\omega_o\alpha) ,$$
$$m\dot{v}_2^S = R_2 = -k(v_2^S + h\dot{\alpha} + r\omega_o\beta) .$$

The third coordinate equation is always complied with because, at least for the purpose of a first approximation, the centre of gravity S does not change its height when the inclinations of the body are small.

For the purpose of continuing the calculation one now introduces the complex variables

(177)
$$\vartheta = \alpha + i\beta \quad ; \quad w = v_1^S + i\,v_2^S$$

The combination of the systems (175) and (176) will then yield the complex differential equations

$$\left.\begin{aligned} A\ddot{\vartheta} + b\dot{\vartheta} - c\vartheta - ikhw &= 0 \;, \\ ikh\dot{\vartheta} + kr\omega_o\vartheta + m\dot{w} + kw &= 0 \;. \end{aligned}\right\} \qquad (178)$$

which contain the abbreviations

$$b = kh^2 - iC\omega_o \;; \quad c = aG + ikhr\omega_o \;.$$

6.2.2. The Stability Conditions for the Toy Top

If we now write

$$\vartheta = \theta e^{\lambda t} \qquad ; \qquad w = We^{\lambda t}$$

(178) yields the characteristic equation

$$\begin{vmatrix} A\lambda^2 + b\lambda - c & -ikh \\ ikh\lambda + kr\,\omega_o & m\lambda + k \end{vmatrix} = 0$$

When this is calculated out and re-ordered, it assumes the form

$$\lambda m\left[A\lambda^2 - iC\omega_o\lambda - aG\right] + \\ + k\left[(A + mh^2)\lambda^2 - i(C + mhr)\omega_o\lambda - aG\right] = 0 \;. \qquad (179)$$

On the basis of this equation one can readily isolate and discuss the limiting cases $k = 0$ (no friction between the body and the base) and $k \to \infty$ (an absolutely rough base, pure rolling). When k has any other arbitrarily chosen values, one can proceed as follows: the system finds itself at the limit of stability when $\lambda = i\omega$ is purely imaginary. But in that case both the expressions in the square brackets of

(179) become real. Since the first of these is once again multiplied by λ, (179) can only be satisfied if both the expressions in the square brackets disappear. In that case we must necessarily have

$$(180) \qquad mh\lambda\,(h\lambda - ir\,\omega_o) = 0$$

as can readily be ascertained by subtracting one expression from the other. But it follows from (180) that one of the roots must be given by

$$\lambda_1 = i\,\frac{r\omega_o}{h} .$$

If this is now substituted in the first expression in square brackets of (179), one obtains the relationship

$$(182) \qquad \omega_o^2\,\frac{r}{h}\left(C - A\,\frac{r}{h}\right) - aG = 0 ,$$

and this contains nothing other than parameters of the system. This relationship holds at the limit of stability; in the stable domain, on the other hand, the expression on the left-hand side of (182) must be positive. This can immediately be seen when one considers the limit of the non-rotating top ($\omega_o = 0$), which can only be stable when $a > 0$ One therefore obtains the inequality

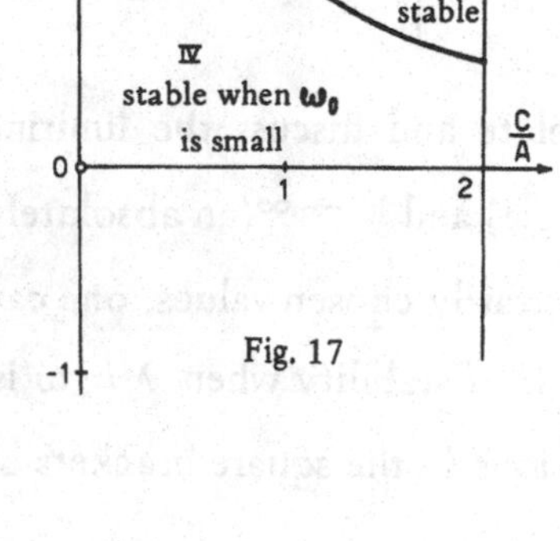

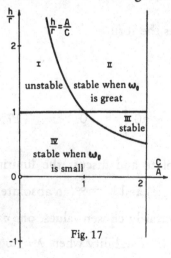

Fig. 17

$$(183) \qquad \omega_o^2\,\frac{r}{h}\left(\frac{C}{A} - \frac{r}{h}\right) > \frac{aG}{A} .$$

as a necessary condition of stability. We can interpret this by considering an $\left(\frac{h}{r}, \frac{C}{A}\right)$-plane (see Fig. 17). Since $0 < C/A < 2$, we need only examine a strip of this plane. This strip is divided into four domains by

the straight line $\frac{h}{r} = 1$ (which corresponds to $a = 0$) and

the hyperbola $\frac{h}{r} = \frac{1}{C/A} = \frac{A}{C}$

and each of these four domains corresponds to a different stability behaviour as follows:

Domain I: $1 < \frac{h}{r} < \frac{A}{C}$. Since $a + r = h$, we have $a > 0$ the left-hand side of (183) is negative, and the motion is therefore always unstable, this quite independently of the magnitude of the rotation ω_o of the point of contact.

Domain II: $1 < \frac{h}{r}; \frac{A}{C} < \frac{h}{r}$. When one introduces the critical value of the rotation

$$\omega_\kappa^2 = \frac{aG}{A \; \frac{r}{h}\left(\frac{C}{A} - \frac{r}{h}\right)} \qquad (184)$$

the motion in Domain II is undoubtedly unstable when $\omega_o < \omega_k$; on the other hand, stability is possible when $\omega_o > \omega_k$.

Domain III: $\frac{A}{C} < \frac{h}{r} < 1$. We now have $a < 0$, and consequently (183) is always satisfied, no matter what the values of ω_o may be; stability is therefore possible.

<u>Domain IV:</u> $\frac{h}{r} < 1$; $\frac{h}{r} < \frac{A}{C}$. (184) shows that the motion in this domain is undoubtedly unstable when $\omega_o > \omega_k$, while stability is to be expected when $\omega_o < \omega_k$.

The critical speed of rotation ω_k depends on the parameters of the system; on the limiting hyperbola $h/r = A/C$ it will tend towards infinity. In the light of what has been said about the stability in the various domains, a fast-spinning top ($\omega_o \gg \omega_k$) will therefore have the stability diagram shown in Fig. 18a; a slow-spinning top ($\omega_o \ll \omega_k$) on the other hand, will have the stability diagram shown in Fig. 18b, this being identical to the diagram for static stability (i.e. $h<r$, and $a<0$).

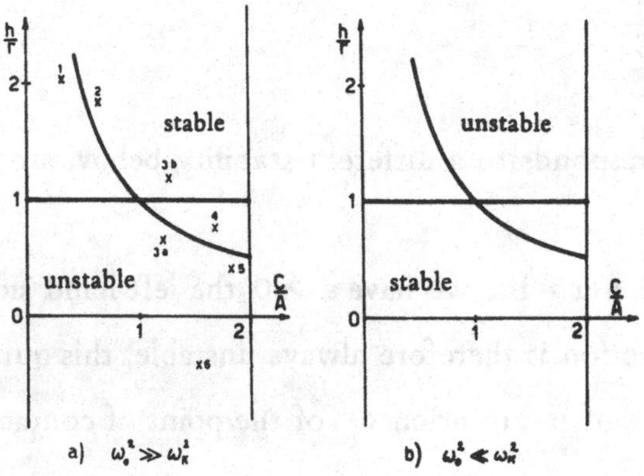

Fig. 18

6.2.3. Conclusions from the Stability Diagram

The stability diagram as obtained in the previous section can be confirmed by means of experiment, and this immediately shows that the assumed friction law reflects the real conditions in a satisfactory manner. Moreover, it should be pointed out that the condition (183) no longer contains the friction coefficient k itself.

The points 1-6 plotted in the stability diagram of Fig. 18a correspond to the various tops whose cross sections are shown in Fig. 19. Top 1 cannot be made to dance, and this no matter how fast it may be caused to spin. Top 2 is

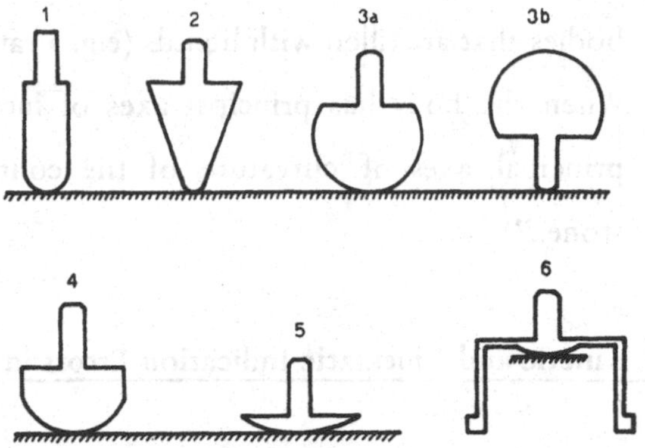

Fig. 19

a well known dancer, always provided that it spins sufficiently fast. Top 4 is stable for all speeds of rotation, while Top 5 is so only when its speed of rotation is very low indeed. Top 3, the so called "tippe-top", is a particularly interesting case. When the top is in its statically stable position (i.e. as shown in 3a) and spinning sufficiently fast, the point representing it in the stability diagram lies in the unstable domain; after being set in motion, the top will therefore leave the position shown in Fig. 19/3a, turn completely upside down and will smoothly rotate on its stem, i.e. in the position shown in Fig. 19/3b. Consequently, the top can rotate stable in this position, always provided that its angular momentum is sufficiently great. It should be noted, however, that the approximate analysis we have made in this section can do no more than to describe the beginning and the end of the turning-over process.

In the case of Top 6 we have $h<0$; crossectional forms of this kind were used, for example, in the well-known Fleuriais horizon. Here it is possible to have instability when the base is flat, but this instability can be avoided when the flat base is replaced by a concave one.

Special effects can be observed in the following cases:

1) When the surface on which the body is rotating does not have a spherical form. An example of this is provided by egg-shaped bodies lying "on their belly", i.e. on their side (e.g. a boiled egg).
2) When the body is not rigid. An example of this type is provided by rotating

bodies that are filled with liquids (e.g. a raw egg).

3) When the body has principal axes of inertia that do not coincide with the principal axes of curvature of the contact surface (e.g. Celtic "wobbling stones").

7. Kinetic and Kinematic Indication Errors in Gyroscopic Instruments

7.1. Oscillation Effects

When oscillations occur in gyroscopic instruments, they can cause certain phenomena that will, at least in part, influence the operation of the instruments themselves. This becomes particularly important in the case of the so-called rectifying effects where, as a result of certain unsymmetries that are inherent in the system, there may occur some unilateral drifts or some unilateral disturbing torques. Disturbances of this kind may be caused by:

1. Effects of the inertia of the gimbals;
2. An intermittent (pulsating) drive of the rotor;
3. Elastic deformations of the structure;
4. Kinematic coupling of the oscillations in the case of gyros with two degrees of freedom.

In the following sections it is proposed to describe a number of typical examples of the first three types of disturbances.

7.1.1. Effects of Gimbal Inertia

Here we have to distinguish between two fundamentally different cases according to whether the instrument housing - and with it the axis of the outer gimbal - remains in a fixed position or can itself perform oscillations. When the instrument housing remains fixed, the gimbal system and the rotor can perform oscillations that are either eigenoscillations (nutations) or oscillations caused by unbalances. Self-excitations caused by control devices can also occur. When the instrument housing is moving, on the other hand, the oscillations always come from outside and are partially transmitted to the gimbal system.

We now propose to treat a particular case of the kinetic drift of a cardanic suspended gyro that is oscillating. For this purpose we shall use the equations of a motion as stated in chapter 4.2, but putting S = O in order to adapt these equations to the case of the astatically supported gyro. In this case we obtain

$$\left.\begin{aligned} &\ddot{\alpha}\left[A\cos^2\beta + (A^A+C^J)\sin^2\beta\right] - \dot{\alpha}\dot{\beta}\,(A^R+A^J-C^J)\sin 2\beta\ + \\ &\qquad + \dot{\beta}C^R\omega_o\cos\beta = 0\ , \\ &\ddot{\beta}B + \dot{\alpha}^2(A^R+A^J-C^J)\sin\beta\cos\beta - \dot{\alpha}C^R\omega_o\cos\beta = 0\,. \end{aligned}\right| \tag{185}$$

These equations have the particular solution $\dot{\alpha} = 0,\ \beta = \beta_o$. This solution states that the axis of symmetry of the rotating gyro can assume and maintain any desired direction with respect to space. For the immediately adjacent motions we can now write

$$\beta = \beta_o + x \tag{186}$$

where x represents a small deviation. After linearizing (185) as far as this magnitude is concerned, one therefore obtains

$$
\begin{aligned}
&\ddot{\alpha} A^{o} + \dot{\beta} C^{R} \omega_{o} \cos\beta_{o} = \dot{\alpha}\dot{x}\,(A^{R} + A^{J} - C^{J})\,\sin 2\beta_{o} + \\
&\quad + x\left[\ddot{\alpha}(A^{R} + A^{J} - C^{J})\,\sin 2\beta_{o} + \dot{\alpha}\dot{x}\,(A^{R} + A^{J} - C^{J})\,2\cos 2\beta_{o} + \right. \\
&\quad \left. + \dot{x} C^{R} \omega_{o} \cos\beta_{o}\right], \\
&\ddot{\beta} B - \dot{\alpha} C^{R} \omega_{o} \cos\beta_{o} = - \dot{\alpha}^{2} (A^{R} + A^{J} - C^{J}) \sin\beta_{o} \cos\beta_{o} - \\
&\quad - x\left[\dot{\alpha}^{2} (A^{R} + A^{J} - C^{J}) \cos 2\beta_{o} + \dot{\alpha} C^{R} \omega_{o} \sin\beta_{o}\right]
\end{aligned}
\tag{187}
$$

In these equations we have used the abbreviation

$$A^{o} = A\cos^{2}\beta_{o} + (A^{A} + C^{J})\sin^{2}\beta_{o} ,$$

and this may be described as the average effective moment of inertia with respect to the axis of the outer gimbal. Moreover, all the non-linear terms on the right-hand sides of equations (187) have been combined.

Equations (187) can be solved in an approximate manner, by means of iteration for example, and the first step then considers only the linear terms. This yields a solution

$$
\begin{aligned}
&\alpha = \alpha_{A} \cos \omega^{N} t , \\
&\beta = \beta_{o} + \alpha_{A} \sqrt{\frac{A^{o}}{B}} \sin \omega^{N} t = \beta_{o} + x
\end{aligned}
\tag{188}
$$

where α_A represents the amplitude of nutation and

$$\omega^N = \frac{C^R \omega_o \cos \beta_o}{\sqrt{A^o\ B}} \tag{189}$$

the frequency of nutation.

In passing on to the second step of the iteration, we now substitute (188) in the expressions on the right-hand sides of equations (187), which will henceforth be designated by the abbreviations R_α and R_β. In view of (188) these expressions contain periodic components with the circular frequencies ω^N, $2\,\omega^N$, and $3\,\omega^N$.

Since these frequencies are generally very great, it will be advantageous to use an average value over the nutation period $T_N = 2\pi/\omega^N$. With this one obtains

$$\left.\begin{aligned} \overline{R_\alpha} &= \frac{1}{T_N}\int_0^{T_N} R_\alpha\, dt = 0 \\ \overline{R_\beta} &= \frac{1}{T_N}\int_0^{T_N} R_\beta\, dt = \frac{1}{2}\alpha_A^2\, \omega^N \sin\beta_o \left[C^R \omega_o \sqrt{\frac{A^o}{B}} - \right. \\ &\quad \left. - (A^R + A^J - C^J)\omega^N \cos\beta_o \right] . \end{aligned}\right| \tag{190}$$

Using (190), it now becomes possible to calculate an average drift speed of the gyro from equations (187), i.e.

(191)
$$\bar{\dot{\alpha}} = -\frac{\alpha_A^2 \, C^R \omega_o \sin\beta_o \, (A^A + C^J)}{2A^o B}$$
$$\bar{\dot{\beta}} = 0$$

The axis of the figure of the gyro therefore drifts in a plane that is normal to the axis of the outer gimbal and does so with an average angular velocity $\bar{\dot{\alpha}}$; this average angular velocity also depends on the average tilt β_o of the inner gimbal.

In deriving equations (191) we presupposed that the normal positions of the geometrical axes of the system of cardanic suspension coincided with the principal axes of the rotor, the inner gimbal and the outer gimbal. However, this is not always the case with gyroscopic instruments. We therefore have to extend our examination to the influence that the hitherto neglected products of inertia of the gimbals exert on the kinetic drift of the instrument. Theory [23] shows that, in place of (191), one must then expect an average speed of drift of the outer gimbal as given by

(192)
$$\bar{\dot{\alpha}} = \frac{\beta_A^2 \, H \, [E^J \cos\beta_o - (A^A + C^J) \sin\beta_o]}{2\,[A^o - 2E^J \cos\beta_o \sin\beta_o]}$$

When $E^J = 0$, and considering that $H = C^R \omega_o$ and that $\beta_A = \alpha_A \sqrt{A^o/B}$, it can be seen that (192) reduces to (191). One should also note the fact that only the product of inertia E^J of the inner gimbal exerts any influence. Both D^J and F^J, as well as the products of inertia of the outer gimbal, are missing from the final result. It can also be seen from equation (192) that a kinetic drift is possible even when $\beta_o = 0$. If one wants to avoid this, it become necessary to balance the inner gimbal in order to ensure the condition $E^J = 0$.

When the rotor is unbalanced, the cardanic suspended gyro will become subject to externally excited oscillations that have the frequency with which the gyro rotates. The amplitudes and the phase angles of these oscillations can be calculated in the well-known manner, although it should be pointed out that resonance effects can lead to particularly great oscillation amplitudes. In the case of these oscillations, once again, it is possible to obtain averages over a single period and thus to determine the unilateral torques that lead to kinetic drifts [23].

The same applies as regards self-excited oscillations of the gimbal system. In all these cases the resulting drift speeds are found to be proportional to the square of the amplitude of the oscillations. On the other hand, the various types of oscillations depend in different ways on the moments of inertia and the products of inertia of the gimbals, and this is also true as regards the influence of the tilt angles β_0 and the angular momentum H.

The drifts discussed above can have very negative effects when instruments are adjusted in the laboratory and are subsequently used in satellites and space ships under no-gravity conditions. The laboratory adjustement of the instruments is based on a careful compensation of the centre of gravity, this being done in such a way as to ensure that their drift speed relative to a reference system that remains fixed with respect to the fixed stars becomes zero. It may therefore happen that a kinetically conditioned drift speed becomes compensated by a gravity torque produced in the course of this balancing process. Since this gravity torque will no longer come into play in the operating no-gravity conditions that exist in space, or may assume false values during the phase when a space ship is being propelled by its own engines, the instruments can become subject to errors that cannot be observed during laboratory experiments.

In all the effects discussed up to the present we have assumed that the rotor itself participates in the oscillations. On the other hand, when we are concerned with oscillations that are transmitted by the instrument housing or frame, the rotor axis can be assumed to be non-oscillating. If any drift will occur at all, it will take place so slowly that the rotor axis can be regarded as remaining in a fixed position with respect to space for the entire duration of an oscillation period. In that case, however, it becomes possible to calculate the reaction torques of the gimbals from the known motions of the gimbals themselves. From the resulting torque that the inner gimbal transmits to the rotor, this torque being again averaged over one period, it becomes possible to determine the average drift speed of the rotor axis.

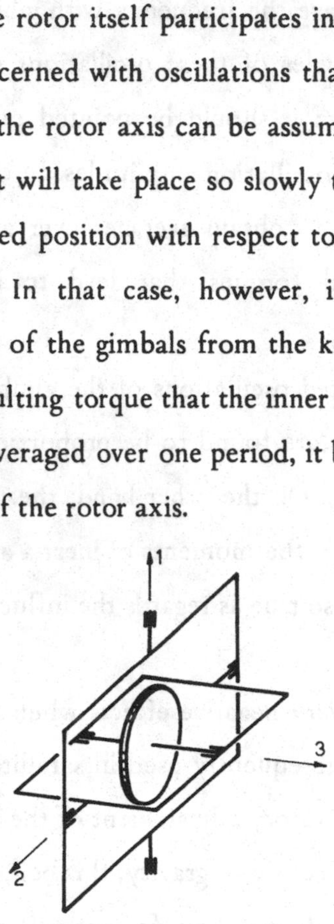

Fig. 20

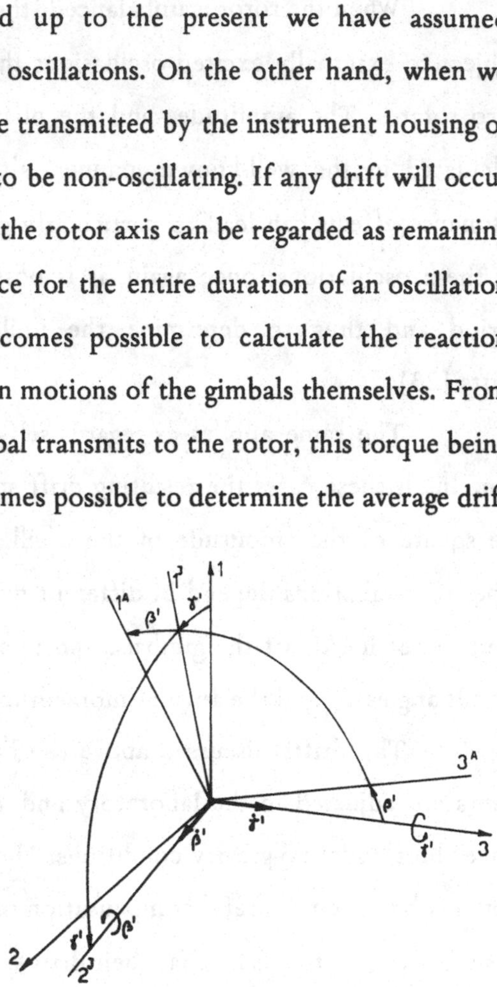

Fig. 21

Let us use the system of axes indicated in Fig. 20. The 3-axis of this system coincides with the direction of the rotor axis, and this latter remains fixed with respect to space. In that case the inner gimbal can rotate only about the 3-axis; let the appropriate angle through which it rotates be designated by γ'. If β' is now used to designate the angle through which the outer gimbal rotates relative to the inner gimbal, then the vectors of the angular velocities (see Fig. 21) will be given by

$$\left.\begin{aligned} \omega_i^J &= (0,\ 0,\ \dot{\gamma}'), \\ \omega_i^A &= (-\dot{\gamma}'\sin\beta',\ \dot{\beta}',\ \dot{\gamma}'\cos\beta'). \end{aligned}\right\} \quad (193)$$

The equations of motion of the inner and outer gimbal, each referred to a system of axes that remains fixed with respect to the gimbal under consideration, can now be written in the Eulerian form and are given by

$$\left.\begin{aligned} \dot{H}_i^J + \epsilon_{ijk}\,\omega_j^J H_k^J &= M_i^J = M_i^{JR} + M_i^{JA}\,, \\ \dot{H}_i^A + \epsilon_{ijk}\,\omega_j^A H_k^A &= M_i^A = M_i^{AJ} + M_i^{AG} \end{aligned}\right\} \quad (194)$$

When the geometric axes of the system of cardanic suspension do not coincide with the principal axes of the gimbals, we will also have the following vectors of angular momentum:

$$\left.\begin{aligned} H_i^J &= \theta_{ij}^J\,\omega_j^J = \begin{bmatrix} -\,E^J\dot{\gamma}' \\ -\,D^J\dot{\gamma}' \\ C^J\dot{\gamma}' \end{bmatrix} \\ H_i^A &= \theta_{ij}^A\,\omega_j^A = \begin{bmatrix} -\,A^A\sin\beta'\dot{\gamma}' - F^A\dot{\beta}' - E^A\cos\beta'\dot{\gamma}' \\ F^A\sin\beta'\dot{\gamma}' + B^A\dot{\beta}' - D^A\cos\beta'\dot{\gamma}' \\ E^A\sin\beta'\dot{\gamma}' - D^A\dot{\beta}' + C^A\cos\beta'\dot{\gamma}' \end{bmatrix} \end{aligned}\right\} \quad (195)$$

As regards the torques that occur in equations (194), the first of the upper indices indicates the body on which this torque is acting and, consequently, it also indicates the coordinate system in which the calculation is being performed. The second of the indices indicates the origin of the torque. Therefore, by way of example, M^{AG} refers to the torque that the instrument housing exerts on the outer gimbal.

Since the equations (194) are valid only in the reference system that remains fixed with respect to the body to which they refer, we shall have to apply the transformation

$$(196) \qquad M_j^{JA} = - a_{ij}^{\beta} M_i^{AJ} = - \begin{bmatrix} \cos\beta' & 0 & \sin\beta' \\ 0 & 1 & 0 \\ -\sin\beta' & 0 & \cos\beta' \end{bmatrix} \begin{bmatrix} M_1^{AJ} \\ 0 \\ M_3^{AJ} \end{bmatrix}$$

Accordingly, the torque exerted on the rotor in the 1,2,3–system, which remains fixed with respect to space, will then be obtained in the form

$$(197) \qquad M_j^{RJ} = - a_{ij}^{\gamma} M_i^{JR} = - \begin{bmatrix} \cos\gamma' & -\sin\gamma' & 0 \\ \sin\gamma' & \cos\gamma' & 0 \\ 0 & 0 & 1 \end{bmatrix} \begin{bmatrix} M_1^{JR} \\ M_2^{JR} \\ 0 \end{bmatrix} .$$

In all the above we have tacitly assumed that the bearings of the axes do not have any friction, and consequently we have

$$M_1^{AG} = M_2^{JA} = M_3^{JR} = 0 .$$

With a view to determining the unknown torque M_i^{RJ} we can now use the second of equations (194) to calculate the magnitude of M_i^{AJ}. After applying the transformation (196) and substituting in the first equation of (194) one obtains an expression for M_i^{JR}. With the help of (197) M_i^{RJ} is then brought into the form

$$M_j^{RJ} = - a_{ij}^{\gamma} \Big[a_{ki}^{\beta} \left(- M_k^{AG} + \dot{H}_k^{A} + \epsilon_{klm} \omega_l^{A} H_m^{A} \right) + + \dot{H}_i^{J} + \epsilon_{ikl} \omega_k^{J} H_l^{J} \Big] . \tag{198}$$

We now introduce the functions $\beta'(t)$ and $\gamma'(t)$ into the above expression, and then average over the oscillation period T. We thus obtain

$$\overline{M_i^{RJ}} = \frac{1}{T} \int_0^T M_i^{RJ} \, dt \quad . \tag{199}$$

From this there follow the average drift speeds for the rotor axis, which is now no longer being regarded as fixed with respect to space, i.e.

$$\overline{\dot{\alpha}} \approx - \frac{\overline{M_2^{RJ}}}{H \cos \beta_o} \quad ; \quad \overline{\dot{\beta}} \approx \frac{\overline{M_1^{RJ}}}{H \cos \beta_o} \quad . \tag{200}$$

Although the actual calculation of (198), and consequently also that of (200), can be rather toilsome in general cases, a large number of components will disappear when an average value is formed in accordance with (199).

7.1.2. Pulsating Drive of the Rotor

The drive of the rotor is governed by the angular velocity relative to the stator; in the case of a cardanic suspended gyro it will therefore depend on $\dot{\gamma}$. In the case of equilibrium we have $\dot{\gamma} = \omega_0$ = const. However, in the case of an oscillating gyro the absolute angular velocity is given by

$$\omega_3' = \dot{\gamma} + \dot{\alpha} \sin\beta \approx \text{const.} \tag{201}$$

For this reason the relative angular velocity $\dot{\gamma}$ becomes smaller when $\dot{\alpha}\sin\beta > 0$. Consequently the rotor will become subject to an additional driving torque $\Delta M_3'$ whose magnitude will be given by

$$\Delta M_3' = - k \Delta\dot{\gamma} \approx k\Delta(\dot{\alpha} \sin\beta) = k \dot{\alpha}\sin\beta \quad , \tag{202}$$

It should be noted that the constant k in the above expression can assume rather considerable values in the case of synchronous gyros. The countertorque balancing the additional driving torque (202) acts on the stator (inner gimbal) and gives rise to a component ΔM_1 acting in the direction of the axis of the outer gimbal. The magnitude of this component is given by

$$\Delta M_1 = - \Delta M_3' \sin\beta = - k \sin^2\beta \dot{\alpha} \tag{203}$$

In the case of nutational oscillations in accordance with (188), equation (203) assumes the form

$$\begin{aligned} \Delta M_1 &= - k \dot{\alpha} (\sin^2\beta_0 + x \sin 2\beta_0) = k \sin^2\beta_0 \alpha_A \overset{N}{\omega} \sin \overset{N}{\omega} t + \\ &+ k \sin 2\beta_0 \alpha_A^2 \sqrt{\frac{A^0}{B}} \overset{N}{\omega} \sin^2 \overset{N}{\omega} t. \end{aligned} \tag{204}$$

This torque does not disappear when it is averaged over the period of a single nutation; rather, there still remains an average value given by

$$\overline{\Delta M_1} = \frac{\omega^N}{2\pi} \int_0^{T_N} \Delta M_1 \, dt = \frac{1}{2} \omega^N k \sin 2\beta_o \alpha_A^2 \sqrt{\frac{A^o}{B}} .$$

With the known values

$$\bar{\dot{\beta}} = \frac{\overline{\Delta M_1}}{H \cos\beta_o} \quad \text{and} \quad \omega^N = \frac{H \cos\beta_o}{\sqrt{A^o B}}$$

it now follows that the drift speed of the inner gimbal is given by

$$\bar{\dot{\beta}} = \frac{k\,\alpha_A^2 \sin 2\beta_o}{2B} . \tag{205}$$

If the inner gimbal already finds itself in a tilted position, the tilt will thus be increased.

7.1.3. The Effects of Elastic Deformations

The constructional elements used to support a gyro rotor, and among these we find such features as shafts, frames, suspensions and housings, are never completely rigid. Under the influence of external stresses, as may readily be induced by accelerations for example, there will always be deformations. These deformations, once again, can give rise to unilateral disturbing torques that will lead to undesired indication errors or drifts. In this section we shall therefore examine the case of deformations in greater detail, but shall limit our considerations to deformations of the elastic kind.

Whereas we always presupposed rotatory oscillations when considering the oscillation phenomena discussed in the preceding sections, we shall now suppose the housing to become subject to translatory oscillations. If $a_i = (a_1, a_2, a_3)$ represents the acceleration vector of these oscillations, the magnitude of the reaction force acting at the centre of gravity of the system will be given by

$$F_i^R = - m\, a_i \tag{206}$$

On account of the elastic deformations occurring in the structure, however, the centre of gravity of the system is deflected from its rest position O through a vector r_i, and the magnitude of this vector can be calculated from

$$F_i^R = c_{ij}\, r_j \tag{207}$$

Here it has to be presupposed that these deflections can be statically determined. This assumption is undoubtedly justified when the frequency of the externally induced periodic motion of the housing is substantially smaller than the frequency of the eigenoscillation caused by the elastic deformations of the structure. In most cases, indeed, this condition is satisfied. When the principal axes of the tensor of constraint c_{ij} coincide with the reference axes of the system, then we have

$$F_i^R = \begin{bmatrix} c_1\, r_1 \\ c_2\, r_2 \\ c_3\, r_3 \end{bmatrix} \quad \text{and} \quad r_i = \begin{bmatrix} F_1^R/c_1 \\ F_2^R/c_2 \\ F_3^R/c_3 \end{bmatrix} \tag{208}$$

The reaction force F_i^R acting at the displaced centre of gravity S now produces a torque given by

$$M_i = \epsilon_{ijk} r_j F_k \ . \tag{209}$$

about the point 0, i.e. the point that represents the rest position of the centre of gravity. Using (208) and (206), it therefore follows that

$$M_i = \begin{bmatrix} F_2 F_3\left(\frac{1}{c_2} - \frac{1}{c_3}\right) \\ F_3 F_1\left(\frac{1}{c_3} - \frac{1}{c_1}\right) \\ F_1 F_2\left(\frac{1}{c_1} - \frac{1}{c_2}\right) \end{bmatrix} = m^2 \begin{bmatrix} a_2 a_3 \frac{c_3 - c_2}{c_2 c_3} \\ a_3 a_1 \frac{c_1 - c_3}{c_3 c_1} \\ a_1 a_2 \frac{c_2 - c_1}{c_1 c_2} \end{bmatrix} \ . \tag{210}$$

From the above it can be seen that no such torque can appear in the presence of isoelastic bearings, because in that case we will have $c_1 = c_2 = c_3$. One therefore speaks of the anisoelastic effects of the bearings.

By way of example, we shall now examine the extremely important case of a constant acceleration (earth acceleration!). When the vector of the acceleration due to gravity lies, for example, in the 2,3–plane and forms an angle φ with the 2–plane, then we have

$$a_1 = 0 \ ; \ a_2 = - g \cos \varphi \ ; \ a_3 = - g \sin \varphi \ .$$

In that case it follows from (210) that

$$(211) \qquad M_1 = \frac{m^2 g^2 (c_3 - c_2) \sin 2\varphi}{2 c_2 c_3} \; ; \qquad M_2 = M_3 = 0 \; .$$

The torque produced by gravity thus disappears in the principal directions $\varphi = 0, \pi/2, \pi, 3\pi/2$. It is proportional to the difference $(c_3 - c_2)$ of the constraint coefficients and is also proportional to the square of the acceleration (g^2-effect!). When the direction of the acceleration vector remains constant with respect to the instrument, there is no difficulty in compensating this torque. However, when gyroscopic instruments are tested on turntables, it is common practice to investigate different positions of the instrument with respect to the gravitational field. Given the consequent variations of the angle φ, the previously discussed anisoelastic effect of the bearings will then make itself felt as a disturbance of the double frequency of the changes in φ.

As a further example, let us consider a linear oscillation of the instrument housing whose amplitude is $s_{io} = (s_{10}, s_{20}, s_{30})$ and whose frequency is Ω. In that case we will have

$$(212) \qquad \alpha_i = \Omega^2 s_{io} \cos \Omega t .$$

If we now substitute this in (210), we will obtain a torque that is proportional to $\cos^2 \Omega t$. It therefore follows that when we average this torque over a period T, i.e.

$$(213) \qquad \overline{M_i} = \frac{1}{T} \int_0^T M_i \, dt$$

there will remain a residual torque given by

$$\overline{M}_i = \frac{1}{2} m^2 \Omega^4 \begin{bmatrix} s_{20}\, s_{30} \dfrac{c_3 - c_2}{c_2\, c_3} \\ s_{30}\, s_{10} \dfrac{c_1 - c_3}{c_3\, c_1} \\ s_{10}\, s_{20} \dfrac{c_2 - c_1}{c_1\, c_2} \end{bmatrix} \qquad (214)$$

One can now convince oneself quite readily that no average anisoelastic torques can be produced in the case of circular translatory oscillations, say oscillations having the form

$$\left.\begin{aligned} a_2 &= s\,\Omega^2 \cos \Omega t \;, \\ a_3 &= s\,\Omega^2 \sin \Omega t \end{aligned}\right| \qquad (215)$$

The torques (211) or (214) will cause drift phenomena in free gyros, while in constrained gyros they will produce indication errors. Both these phenomena can be calculated by means of already known procedures.

7.2. Kinematic Indication Errors in Cardanic Suspended Positional Gyros

Positional gyros are built into a moving carrier (ship, aircraft, or satellite) in order to be able to determine the rotational motions of the carrier from the changes that occur in the gimbal angles. The relative orientation of a reference system $1^T, 2^T, 3^T$ (which remains fixed with respect to the carrier) and a system $1^K, 2^K, 3^K$ (which remains fixed with respect to the cardanic suspended gyro) becomes of importance in this connection. Without developing the appropriate theory in complete detail, we shall here deal with some of the basic concepts that govern an analytical determination of the kinematic indication errors that may occur.

Let the position of the carrier be characterized by the angles that are usually adopted in flight mechanics, i.e. ψ (course or yaw), ϑ (inclination or pitch), and φ (banking or roll). Similarly, let the state of the gyro system be determined by means of the gimbal angles α and β. We are therefore interested in the functions

$$\alpha = \alpha\ (\psi, \vartheta, \varphi) \quad \text{and} \quad \beta = \beta\ (\psi, \vartheta, \varphi) \ . \tag{216}$$

The calculation of these functions can be based on the assumption that the direction of the rotor axis remains fixed with respect to space, i.e. that the gyro does not become subject to any drift phenomena.

The system $1^T, 2^T, 3^T$, which remains fixed with respect to the carrier itself, can be brought into any desired position by consecutively carrying out three simple rotations about the axes of the inertial system 1, 2, 3, these axes being kept fixed in each case (see Fig. 22). A position vector r_i can then be referred to the various coordinate systems by means of the following transformation formulas:

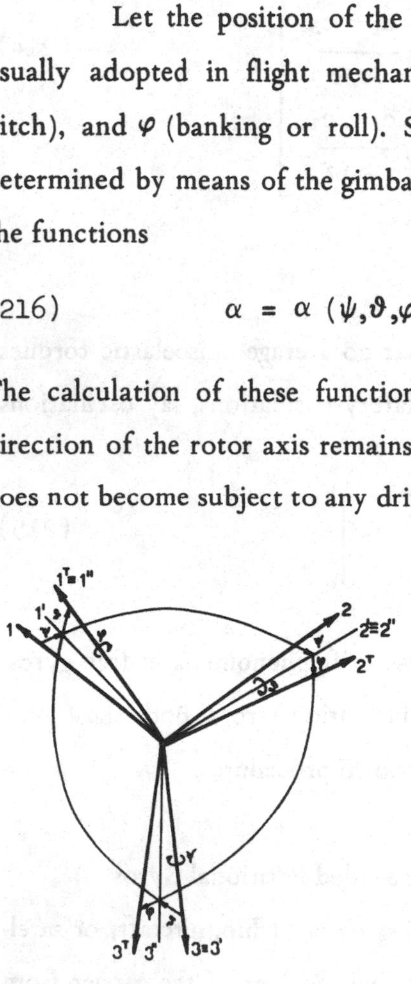

Fig. 22

$$r_i = a^{\psi}_{ij}\, r'_j = a^{\psi}_{ij}\, a^{\vartheta}_{jk}\, r''_k = a^{\psi}_{ij}\, a^{\vartheta}_{jk}\, a^{\varphi}_{kl}\, r^T_l = a^T_{ij}\, r^T_j \ . \tag{217}$$

The terms a^{ψ}_{ij}, a^{ϑ}_{jk}, a^{φ}_{kl} in the above formulae represent transformation matrices that characterize simple rotations through the angles ψ, ϑ, φ. On the other hand, a position vector r^J_i defined in the reference system that remains fixed with respect to the inner gimbal can be transformed into the system that remains fixed with respect to the carrier via the reference systems that respectively remain fixed with respect to the outer gimbal (A) and the instrument housing (G). For this transformation we have

$$\begin{aligned} r^A_i &= a^{\beta}_{ij}\, r^J_j \ , \\ r^G_i &= a^{\alpha}_{ij}\, r^A_j = a^{\alpha}_{ij}\, a^{\beta}_{jk}\, r^J_k \ ; \\ r^T_i &= a^E_{ij}\, r^G_j = a^E_{ij}\, a^{\alpha}_{jk}\, a^{\beta}_{kl}\, r^J_l \ . \end{aligned} \tag{218}$$

The term a^E_{ij} in the above transformation represents the orientation matrix that can be obtained, for example, from the matrix a^T_{ij} by replacing the angles ψ, ϑ, φ with the angles $\psi_o, \vartheta_o, \varphi_o$, these latter being the angles that characterize the position in which the gyroscopic instrument is built in with respect to the carrier. The terms a^{α}_{jk} and a^{β}_{kl} once again represent matrices for simple rotations. The relationship between a vector in the system of the inner gimbal and in the system that remains fixed with respect to space can then be found as follows

$$r_i = a^T_{ij}\, a^E_{jk}\, a^{\alpha}_{kl}\, a^{\beta}_{lm}\, r^J_m \ . \tag{219}$$

If one now takes into consideration the fact that the unit vector in the direction of the rotor axis, i.e. $r^{Jo}_i = (0,0,1)$, remains fixed with respect to space and that, on account of $a^{\alpha}_{ij}(0)=a^{\beta}_{ij}(0)=\delta_{ij}$, the original position of the rotor axis in the system that remains fixed with respect to space is given by

$$r^{o}_{i} = a^{E}_{ij}\, r^{Jo}_{j}$$

one now finds from (219) that any given motions of the carrier can be represented by the relationship

$$a^{E}_{ij}\, r^{Jo}_{j} = a^{T}_{ij}\, a^{E}_{jk}\, a^{\alpha}_{kl}\, a^{\beta}_{lm}\, r^{Jo}_{m} \,. \tag{220}$$

For the purpose of a readier interpretation of this relationship it will be advantageous to transform (220) through multiplication by a^{T}_{ji}. One thus obtains

$$a^{T}_{ji}\, a^{E}_{jk}\, r^{Jo}_{k} = a^{E}_{ij}\, r^{Go}_{j} \,, \tag{221}$$

where

$$r^{Go}_{i} = a^{\alpha}_{ij}\, a^{\beta}_{jk}\, r^{Jo}_{k} = \begin{bmatrix} \sin\beta \\ -\sin\alpha\,\cos\beta \\ \cos\alpha\,\cos\beta \end{bmatrix} \tag{222}$$

represents the unit vector in the direction of the rotor axis taken in the reference system that remains fixed with respect to the instrument frame (housing). Its coordinates now depend only on the angles α and β. Since the position angles ψ,ϑ,φ are contained in a^{T}_{ij} and the building–in angles $\psi_o,\vartheta_o,\varphi_o$ are contained in a^{E}_{ij}, it is now possible to calculate the looked–for functions

$$\alpha = \alpha\,(\psi,\vartheta,\varphi,\psi_o,\vartheta_o,\varphi_o) \quad \text{and} \quad \beta = \beta\,(\psi,\vartheta,\varphi,\psi_o,\vartheta_o,\varphi_o) \tag{223}$$

from (221) for any given position in which the instrument may be built into the carrier.

In special cases, moreover, these calculations can also take account of any tilt β_0 that may characterize the position of the inner gimbal with respect to the outer gimbal in the normal working position of the instrument. In that case β_0 becomes an additional parameter in (223).

8. The Tuning of Gyroscopic Instruments

Ever since the fundamental work on this subject by Schuler [24], it has been known that certain navigational instruments, including gravity pendulums, gyro pendulums and gyro compasses, lend themselves to particularly successful tuning when the curvature of the earth is taken into consideration. In this way the instruments can be made to become insensitive to the disturbances that are caused as the result of the accelerations of the carriers along the surface of the earth. The requirement according to which these instruments have to be tuned to an oscillation period of 84.3 minutes (Schuler period) is often referred to as the Schuler principle. In spite of the numerous investigations that have been made as regards the contents and the limits of this principle, the relevant explanations given in many textbooks are rather unsatisfactory. In this chapter it is therefore proposed to use two examples to illustrate some of the particular difficulties that are met in the application of the Schuler principle.

8.1. Pendulum and Gyro Pendulum

Let us assume that we have a gyro vertical (gyro pendulum or gyro–horizon), and that this is made up of a symmetrical rotor R contained in an equally symmetrical rotor housing G (see Fig. 23).

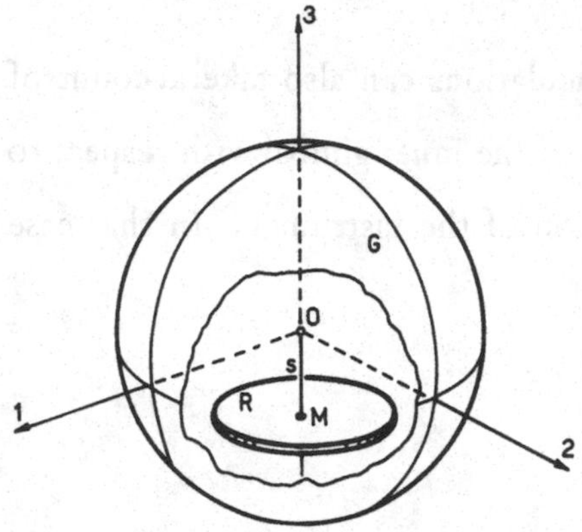

Fig. 23

Let the principal moments of inertia be $A^R = B^R$, C^R and $A^G = B^G$, C^G so that $A = A^R + A^G = B^R + B^G$ and $C = C^R + C^G$. Let us further assume that the rotor housing, which may for example have the form of a spherical float, has a fixed point 0 (point of suspension) and that the position of this point remains invariant with respect to the carrier. We shall also assume the joint centre of gravity M to lie on the 3–axis, at a distance s from 0. The friction and driving torques about the rotor axis will be assumed to be in equilibrium, and the damping torques acting on the spherical float will be neglected.

Let us now suppose that the point of suspension 0 of the gyro pilot is moved in any desired manner along the surface of the earth, this latter being here assumed to be spherical and to have a radius R. The velocity of this motion and the velocity due to the rotation of the earth will be combined into a single velocity vector v_i. In a reference system with the horizontal axes 1 and 2 and a vertical axis 3 this velocity vector will have the components $v_i = (v_1, v_2, 0)$. The reference system must therefore perform a rotation

$$\Omega_i = \left(-\frac{v_2}{R}, \frac{v_1}{R}, 0 \right) \tag{224}$$

if the 3–axis is to remain vertical. In this reference system we can now derive the approximate equations of motion by applying the theorem of angular momentum in the form

$$\frac{dH_i}{dt} + \epsilon_{ijk}\Omega_j H_k = M_i \tag{225}$$

together with (224) and

$$H_i = \left[A\left(\dot{\alpha} - \frac{v_2}{R}\right) + H\beta \ ; \ A\left(\dot{\beta} + \frac{v_1}{R}\right) - H\alpha \ ; \ H \right] \tag{226}$$

In doing so we must presuppose that the gimbal angles α and β remain small, i.e. that the rotor axis 3 comes to deviate only very slightly from the vertical. The magnitude $H = H_o$ represents the constant component of the angular momentum in the direction of the rotor axis.

If we now consider the components due to the acceleration of the point of suspension 0, and also the torques of the force of gravity and of the gravitational gradient, as the external torques in equation (225), we will find the equations of motion to be given by

$$\left.\begin{aligned} A\left(\ddot{\alpha} - \frac{\dot{v}_2}{R}\right) + H\dot{\beta} + \frac{H}{R}v_1 &= M_1 = -\,ms\dot{v}_2 - \left[msg - \frac{3g}{R}(C-A)\right]\alpha \\ A\left(\ddot{\beta} + \frac{\dot{v}_1}{R}\right) - H\dot{\alpha} + \frac{H}{R}v_2 &= M_2 = ms\dot{v}_1 - \left[msg - \frac{3g}{R}(C-A)\right]\beta \ . \end{aligned}\right\} \tag{227}$$

Combining these by means of the complex variables

(228) $$z = \alpha + i\beta \quad \text{and} \quad w = v_1 + i\,v_2$$

and introducing the abbreviation

(229) $$k = mgs - \frac{3g}{R}(C-A)$$

we now obtain

(230) $$A\ddot{z} - iH\dot{z} + kz = i\left(ms - \frac{A}{R}\right)\dot{w} - \frac{H}{R}w = f(t)\ .$$

On the right–hand side of this equation we have a function of time $f(t)$ that depends both on the speed w and the acceleration $\dot{w}$. In the following remarks we shall not make any kind of limiting assumption as regards this function. The point of suspension 0 can therefore perform any desired kind of motion along the surface of the earth.

Before passing on to examine the general solution of equation (230), we propose to take a look at a special solution that is of particular interest. If the rotor is maintained in a fixed position with respect to the instrument housing, then we will have $H = 0$ and the gyro pendulum becomes transformed into a simple pendulum. If the position of the centre of gravity of this simple pendulum is now chosen in such a way as to make

(231) $$s = \frac{A}{mR}\ .$$

then the right–hand side of equation (230) will disappear. But in that case $z \equiv 0$ will become a particular solution. It indicates that the 3–axis of the simple pendulum always remains in a vertical position, and this quite independently of any motions that its point of suspension may perform along the surface of the earth. We would thus seem to have an ideal vertical plumbline that will remain perfectly aligned along the vertical even when it is installed in a moving vehicle. Schuler was the first to draw attention to this possibility. However, the practical realization of such a perfect plumbline fails when it comes to complying with the tuning condition (231). This condition states that the reduced length of the pendulum must be equal to the radius R of the earth. The period of oscillation of a pendulum tuned in this manner is then found to be given by

$$T = 2\pi\sqrt{\frac{A}{k}} = 2\pi\sqrt{\frac{R}{g}}\sqrt{\frac{A}{4A-3C}} \,. \tag{232}$$

In the case of a rod–shaped pendulum, where $C = 0$, the numerical value of this period of oscillation is obtained as $T = 42.2$ minutes; in the case of a pendulum whose ellipsoid of inertia has a spherical form one obtains the Schuler period, i.e. $T = 84.4$ minutes; lastly, in the case of a pendulum where $4A = 3C$, the period $T \to \infty$. When $3C > 4A$, as would be the case with a strongly flattened pendulum, the equilibrium position $z = 0$ becomes unstable. Quantitatively false results are obtained when the component due to the gravitational gradient in (227) is neglected: indeed, in that case one will obtain a period of oscillation of 84.4 minutes no matter what the shape of the pendulum may be.

The general solution of equation (230) can now be constructed from the particular solutions for the homogeneous equation, a process with which we are already familiar. One begins by noting that the homogeneous equation yields the two eigenfrequencies

(233)
$$\left.\begin{matrix}\omega^N \\ \omega^P\end{matrix}\right\} = \frac{H}{2A}\left[1 \pm \sqrt{1 + \frac{4Ak}{H^2}}\right]$$

The appropriate partial solutions

(234)
$$z_1 = Z_1 e^{i\omega^N t} \quad \text{and} \quad z_2 = Z_2 e^{i\omega^P t}$$

form a fundamental system from which the general solution of equation (230), i.e. for $f(t) \neq 0$, is found as

$$z = z_1\left[1 + \int_0^t \frac{z_2 f(t)}{A(z_2 \dot{z}_1 - z_1 \dot{z}_2)}\, dt\right] + z_2\left[1 + \int_0^t \frac{z_1 f(t)}{A(z_1 \dot{z}_2 - z_2 \dot{z}_1)}\, dt\right],$$

When this is calculated out with the help of (234), one obtains

(235)
$$z = e^{i\omega^N t}\left[Z_1 + \frac{i}{A(\omega^P - \omega^N)}\int_0^t f(t)\, e^{-i\omega^N t}\, dt\right] + e^{i\omega^P t}\left[Z_2 + \frac{i}{A(\omega^N - \omega^P)}\int_0^t f(t)\, e^{-i\omega^P t}\, dt\right].$$

If in the integrals that appear in (235) we now replace the function f(t) by the expression provided by (230), it becomes possible for the components depending on the speed w to be transformed by means of partial integration, and in this manner we obtain them as

$$\int_0^t w\, e^{-i\omega t}\, dt = \frac{i}{\omega}\left[w\, e^{-i\omega t} - w_o \right] - \frac{i}{\omega}\int_0^t \dot{w}\, e^{-i\omega t}\, dt \ . \quad (236)$$

Using this result together with the new constants

$$\overset{\star}{Z}_1 = Z_1 + \frac{H w_o}{A R \omega^N (\omega^N - \omega^P)} \ ;$$

$$\overset{\star}{Z}_2 = Z_2 - \frac{H w_o}{A R \omega^P (\omega^N - \omega^P)}$$

the general solution (235) can be brought into the form

$$z = \left[\overset{\star}{Z}_1 e^{i\omega^N t} + \overset{\star}{Z}_2 e^{i\omega^P t} \right] + \frac{H w}{A R \omega^N \omega^P} +$$

$$+ \frac{e^{i\omega^N t}}{A(\omega^N - \omega^P)} \left[ms - \frac{A}{R} + \frac{H}{R\omega^N} \right] \int_0^t \dot{w}\, e^{-i\omega^N t}\, dt \ +$$

$$(237)\qquad + \frac{e^{i\omega^P t}}{A(\omega^P - \omega^N)}\left[\, ms - \frac{A}{R} + \frac{H}{R\omega^P}\,\right]\int_0^t \dot{w}\, e^{-i\omega^P t}\, dt \;.$$

In this solution the two terms in the first square bracket represent the eigenoscillations nutation and precession; the third term characterizes an erroneous indication of the gyro vertical that depends on the speed w and therefore contains both the error due to the rotation of the earth and the error due to the motion of the carrier itself. If one takes into account the speed of the earth's rotation, the geographical latitude, the course bearing and the speed of the carrier, one will experience no difficulty in evaluating each of these components. The last two terms of (237) depend on the acceleration $\dot{w}$; they contain factors that can only be made to disappear simultaneously when $H = 0$ and the tuning condition (231) is fulfilled. With this we have immediately demonstrated that an appropriate tuning of the instrument parameters that will completely eliminate the influence of the accelerations of the instrument carrier exists only in the special case that we have already discussed. When $H \neq 0$, we also have $\omega^N \neq \omega^P$, and it is therefore clear that the two factors in front of the acceleration terms in (237) can never disappear simultaneously. It therefore follows that a gravity–influenced gyro vertical can never be tuned in such a way as to make it completely independent of the various accelerations.

In practical cases one must therefore decide to eliminate either the one or the other of the two disturbance terms in (237) by means of suitable tuning. The actual decision can then be based on the frequency spectrum of the accelerations that are to be expected. If one has to expect accelerations that last for a long period of time, i.e. the kind of acceleration that occurs during ship manoeuvres, one will undoubtedly have

$$\int \dot{w}\, e^{-i\omega^{P} t}\, dt \gg \int \dot{w}\, e^{-i\omega^{N} t}\, dt$$

but the relationship between the numerical values of these terms can be exactly the opposite when the point of suspension becomes subject to horizontal disturbances with a higher frequency. In the first case, therefore, one must tune the gyro pilot in such a way as to make

$$ms - \frac{A}{R} + \frac{H}{R\omega^{P}} = 0 \qquad (238)$$

The value of ω^{P} to be used in the above expression can be the approximate one that applies in the case of a large angular momentum. This approximate value is obtained from (233) as $\omega^{P} = -k/H$. Substituting this in (238) and taking account of (229), we thus obtain the following quadratic equation for s:

$$s^2 m^2 R^2 g - smRg(3C-2A) - \left[H^2 R - 3gA(C-A) \right] = 0.$$

For high values of the angular momentum we will surely have $H^2 R \gg 3ga(C-A)$. We can thus find an approximate solution, given by

$$s \approx \frac{H}{m\sqrt{Rg}} \qquad (239)$$

when we neglect the middle term. Indeed, when we substitute the value of s as given by (239), it will readily be seen that this term is small when compared with the expression $H^2 R$. The value of s as given by (239) can also be seen to be great when compared with the value obtained for the physical pendulum (231) because H will always be much greater than $A\sqrt{g/R}$. On account of (239) we can therefore assume k to be given by

$$k \approx msg \approx H\sqrt{\frac{g}{R}}$$

Using this value of k, we then find that the precession period of a gyro vertical tuned in accordance with (239) is given by

$$T = \frac{2\pi}{|\omega^P|} \approx 2\pi\,\frac{H}{k} \approx 2\pi\sqrt{\frac{R}{g}} = 84{,}4 \text{ min.} \tag{240}$$

On the other hand, when it is desired to eliminate the influences of disturbances with a higher frequency, one has to choose

$$ms - \frac{A}{R} + \frac{H}{R\omega^N} \approx 0 .$$

Using the approximate value $\omega^N \approx H/A$, it immediately follows from the above that $s = 0$. In fact, this type of disturbance can be best eliminated by providing the gyro with astatic supports (bearings).

8.2. Gyro Platforms

When examining the gyro vertical in Section 8.1. above, we found that one of the most important results lays in the realization that, both in the case of the ordinary gravity pendulum and in the case of the gyro pendulum, it is possible to tune the instrument in such a way as to reduce its indication errors to a minimum, this being done by means of a suitable choice of the instrument parameters. This is also true as regards the platforms used in inertial navigation. We shall now use the example of the horizonted platform, an instrument frequently used in navigation on the surface of the earth, to make a more detailed examination of the relationships that exist in this case.

8.2.1. The Tuning of a Synthetic Pendulum

A gravity pendulum is a device capable of performing oscillations, and its properties can be explained by the joint action of forces of inertia and restoring forces (i.e. forces that tend to bring it back to an equilibrium position). However, these properties of inertia and restoring capability can also be produced in an artificial manner in compound systems, an example of this type being illustrated in the sketch of Fig. 24. In the synthetic pendulum there shown P represents a platform that can freely rotate about a fixed horizontal 1–axis passing through its centre of gravity, the axis being supported on bearings. Mounted on top of this platform there is an acceleration measuring device B; when the platform deflects from the horizontal position, this device sends a measuring signal via an amplifier to the control motor, and this latter will then exert an appropriate torque on the axis of rotation. The sign of the signal is chosen in such a way that the control motor will reduce the then existing inclination of the platform.

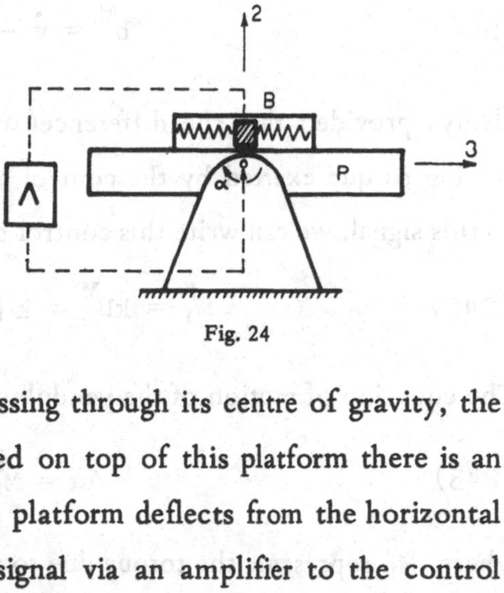

Fig. 24

We shall now examine the behaviour of the synthetic pendulum while it

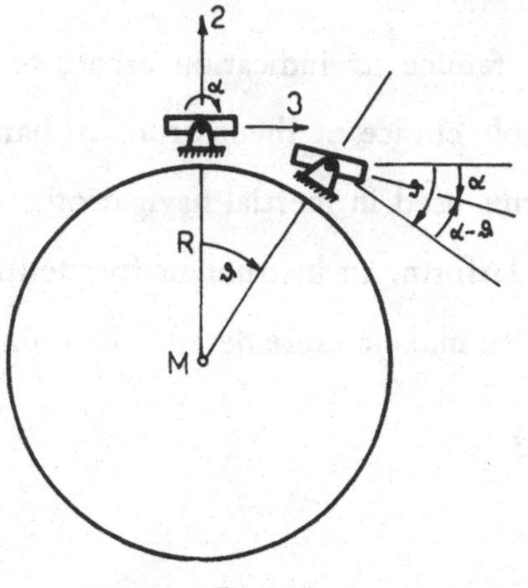

Fig. 25

is being displaced along a great circle on the surface of the earth. This is illustrated in Fig. 25, where ϑ represents the angle through which the straight line joining the centre of the earth M to the carrier of the pendulum rotates as the carrier moves along the great circle. Let α be the absolute angle through which the platform rotates, and $\dot{v} = R\ddot{\vartheta}$ the acceleration of the carrier in the course of its motion. The acceleration measuring device will then record a measured value of

$$b^M = \dot{v} - g\,(\alpha - \vartheta) \tag{241}$$

always provided that the difference $(\alpha - \vartheta)$ remains small. If we assume the restoring torque exerted by the control motor on the platform to be proportional to this signal, we can write this control torque as

$$M_1^K = kb^M = k\,[\,\dot{v} - g\,(\alpha - \vartheta)] \tag{242}$$

The equation of motion of the pendulum can then be formulated as

$$A\ddot{\alpha} = M_1^K + M_1^G \tag{243}$$

where M_1^G represents the torque due to the gravitational gradient. This torque will be acting even when the 1–axis passes through the centre of gravity. For the case of small differences $(\alpha - \vartheta)$, as we have already presupposed, one then obtains

$$M_1^G = 3\,\frac{g}{R}\,(B-C)\,(\alpha - \vartheta). \tag{244}$$

Substituting (242) and (244) in (243), the equation of motion becomes transformed into

$$A\ddot{\alpha} + \left[kg - \frac{3g}{R}(B-C) \right](\alpha - \vartheta) = k\dot{v} . \qquad (245)$$

In view of the fact that $\dot{v} = R\ddot{\vartheta}$, we can now use the tuning condition

$$k = \frac{A}{R} \qquad (246)$$

to bring (245) into the form

$$A(\ddot{\alpha} - \ddot{\vartheta}) + \left[\frac{gA}{R} - \frac{3g}{R}(B-C) \right](\alpha - \vartheta) = 0 \qquad (247)$$

This equation of motion has the particular solution

$$\alpha = \vartheta, \qquad (248)$$

a solution that indicates that the platform P of the pendulum always remains in a horizontal position, and this quite irrespective of the motions carrier may perform along a great circle. With (246) we have therefore found the condition for tuning the pendulum in such a way as to make it completely independent of accelerations. When the initial conditions do not conform to the particular solution (248), or when disturbances come into play, the pendulum, as can readily be seen from (247), will perform undamped oscillations whose period is given by

$$T = 2\pi \sqrt{\frac{R}{g}} \sqrt{\frac{A}{A-3(B-C)}} . \qquad (249)$$

This result corresponds completely to the value (232) obtained in Section 8.1. for the simple symmetrical gravity pendulum. It should be noted once again that, even in the case of the synthetic pendulum we have just examined, the torque caused by the gravitational gradient cannot be neglected. Only in the case of a system having an ellipsoid of inertia that is symmetrical with respect to the 1–axis, i.e. when $B = C$, will we obtain the well–known Schuler period of $T = 2\pi\sqrt{R/g} = 84$ minutes. According to the relationships that exist between the moments of inertia, the real periods of oscillation of the system shown in Fig. 24 may lie between 42.2 minutes and infinity.

8.2.2. The Horizontal Platform Controlled by a Servo Loop

The simple system illustrated in Fig. 24 does not lend itself very readily to technical applications. It is therefore supplemented by the introduction of a servo loop designed to ensure that the speed of rotation of the platform will always comply with the law

$$\dot{\alpha} = \dot{\alpha}_{Soll} = \frac{1}{R}\int b^{M} dt \tag{250}$$

This law states that, given an ideally functioning servo loop, the speed of rotation $\dot{\alpha}$ of the platform will always be the same as the angular velocity $\dot{\vartheta}$ with which the radius connecting the centre of the earth M and the instrument carrier rotates at that moment. The 2–axis will then always coincide with the local direction of a plumbline, always provided that the initial conditions of the instrument were not erroneous.

From (241) and (250) one then obtains the equation of motion of the system as

$$A\ddot{\alpha} = M_1^K = A\ddot{\alpha}_{Soll} = \frac{A}{R} b^M = \frac{A}{R}\left[\dot{v} - g\,(\alpha - \vartheta)\right],$$

or, putting $\dot{v} = R\ddot{\vartheta}$

$$(\ddot{\alpha} - \ddot{\vartheta}) + \frac{g}{R}(\alpha - \vartheta) = 0. \tag{251}$$

This equation, once again, has the desired particular solution $\alpha = \vartheta$. The servo loop law (250) will now ensure that the system will always be characterized by a period of oscillation

$$T = 2\pi\sqrt{\frac{R}{g}} = 84{,}4 \text{ min.} \tag{252}$$

These oscillations will come into being when there are erroneous initial conditions. Unlike the case of the previously discussed pendulum, given a perfectly tuned servo loop, one now obtains the exact Schuler period. The torque due to the gravitational gradient does not appear at all; as a disturbing torque it is completely compensated by the torque produced by the servo loop, indeed, we presupposed that the servo loop was functioning in a perfect manner. The difficulties connected with the support of the platform that are encountered during the practical realization of such a system can therefore be more readily overcome in the case of a system of the type shown in Fig. 26 than in the case of the simpler synthetic pendulum illustrated in Fig. 24.

The servo loop to control the speed of rotation of the platform can be

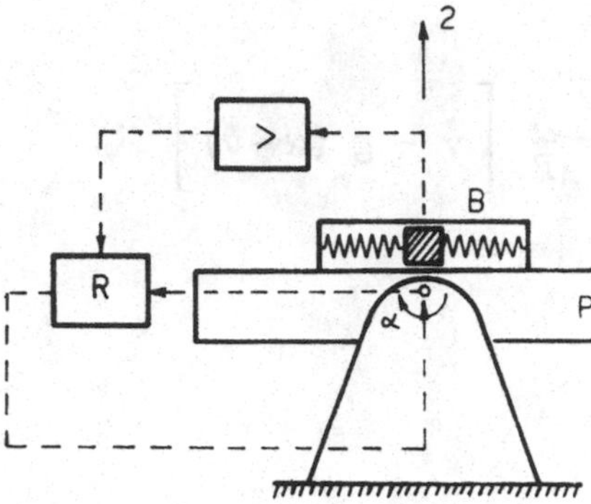

Fig. 26

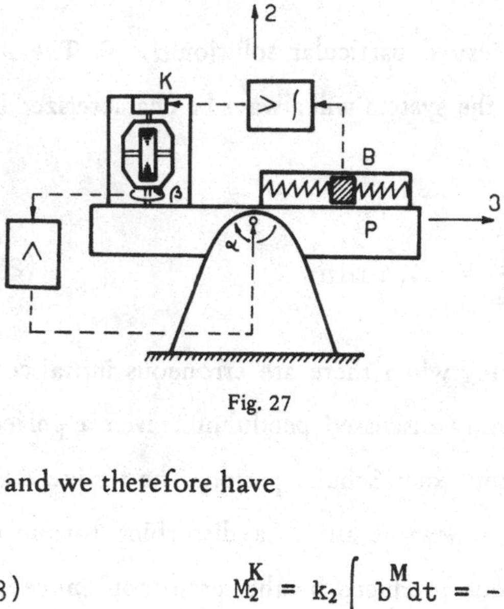

Fig. 27

realized as shown in Fig. 27. The desired control rotation $\dot{\alpha}_{Soll}$ is produced via the torquer on the output axis of a rate gyro, this latter serving as a sensitive control device that determines the speed of rotation of the platform. The servo loop, acting via the amplifier and the control motor, ensures that the average output of the rate gyro will always remain zero.

The correcting torque M_2^K applied to the axis of the platform is made to be proportional to the output value b^M of the acceleration measuring device, and we therefore have

$$(253) \qquad M_2^K = k_2 \int b^M dt = -\frac{H}{R} \int b^M dt \,.$$

In the stationary case, therefore, the requirement (250) is exactly fulfilled.

In real systems, however, there will be deviations from (250). These deviations can be estimated on the basis of a more detailed examination of the equations of motion. In place of the simple equation of motion (251) one must now consider the equilibrium of the torques about both the 1–axis and the gimbal axis of gyro in Fig. 27. One thus obtains

$$\left.\begin{aligned} A\ddot{\alpha} + H\dot{\beta} &= M_1^K + \vartheta = -k_1\beta\,, \\ B^K\ddot{\beta} - H\dot{\alpha} + d\dot{\beta} + c\beta &= M_2^K = -\frac{H}{R}\int b^M dt\,. \end{aligned}\right\} \tag{254}$$

On account of (241) this can be transformed into

$$\left.\begin{aligned} & A\ddot{\alpha} + H\dot{\beta} + k_1\beta = 0\,, \\ & B^K\ddot{\beta} + d\dot{\beta} + c\beta - H\int\left[(\ddot{\alpha} - \ddot{\vartheta}) + \frac{g}{R}(\alpha - \vartheta)\right] dt = 0. \end{aligned}\right\} \tag{255}$$

For any given motion of the carrier along the surface of the earth, $\vartheta = \vartheta(t)$ will always be a known excitation function. It is therefore possible to calculate the externally excited motions of the system (and particularly the behaviour of $\alpha(t)$ from (225). It can be seen from the second of equations (255) that the transient motion of the rate gyro will disturb the ideal solution $\alpha - \vartheta = 0$ that would otherwise be possible as a result of Schuler tuning. But even the servo loop can produce disturbances, as may be seen from the first of equations (255). The possible transient behaviour of the system can be appreciated either by examining the transfer function of (255) or by discussing the roots of the characteristic equation that follows from (255), i.e.

$$\underline{\lambda}^5 AB^K + \underline{\lambda}^4 Ad + \underline{\lambda}^3 (H^2 + Ae) + \underline{\lambda}^2 k_1 H + \underline{\lambda}\,\frac{H^2 g}{R} + \frac{Hgk_1}{R} = 0. \tag{256}$$

Now, it is undoubtedly admissible to assume that H^2 is much greater than Ac. If one further neglects the transient motion of the rate gyro which is equivalent to the disappearance of the two terms containing the highest powers of $\underline{\lambda}(B^K = d = 0)$, one can obtain an approximation to (256) in the form

$$H\left(\underline{\lambda}^2 + \frac{g}{R}\right)(\underline{\lambda}H + k_1) \approx 0. \tag{257}$$

This equation has the roots

$$\underline{\lambda}_{1,2} = \pm\, i\sqrt{\frac{g}{R}}\;; \quad \underline{\lambda}_3 = -\frac{k_1}{H}\,. \tag{258}$$

This result shows that the aperiodic transient process of the servo loop becomes superposed by the 84–minute oscillations of the platform. To this, of course, must be added the disturbances produced by the rate gyro, since these were purposely neglected in the process of deriving the above expressions.

REFERENCES

[1] Duschek, A., and A. Hochrainer: Tensorrechnung in analytischer Darstellung (Tensor Calculus in Analytical Notation), 3 volumes, Vienna, Springer–Verlag, 1960.

[2] Grammel, R.: Der Kreisel, seine Theorie und seine Anwendungen (The Gyro, its Theory and its Applications), 2 volumes, Berlin / Göttingen / Heidelberg, Springer–Verlag, 1950.

[3] Klein, F., and A. Sommerfeld: Über die Theorie des Kreisels (Theory of the Gyro), 4 volumes, Leipzig, Teubner–Verlag, 1910-1922.

[4] Leimanis, E.: The General Problem of the Motion of Coupled Rigid Bodies about a Fixed Point, Berlin / Heidelberg / New York, Springer–Verlag, 1965.

[5] Grammel, R.: Ing. Arch. 22 (1954), pp. 73–97.

[6] Grammel, R.: Ing. Arch. 29 (1960), pp. 153–159.

[7] Weidenhammer, F.: Z. Angew. Math. Mech. 38 (1958), pp. 480-483.

[8] Leipholz, H.: Ing. Arch. 32 (1963), pp. 255-296.

[9] Magnus, K.: Acta Mechanica 2 (1966), pp. 130–143.

[10] Wittenburg, J.: Ing. Arch. 37 (1968), pp. 221–242.

[11] Lurje, A. I.: Arbeiten des Leningrader Polytechnischen Institutes (Papers of the Leningrad Polytechnical Institute), 210 (1960), pp. 7–22.

[12] Roberson, R. E., and J. Wittenburg: Proceedings of the Third IFAC Congress, London, 1966.

[13] Merkin, D. R.: Kreiselsysteme (Gyro Systems), in Russian, Moscow, State Publishing House for Technical and Theoretical Literature, 1956.

[14] Metelizyn, I. I.: Doklady Akad. Nauk SSSR, 86 (1952), pp. 31–34.

[15] Forbat, N.: Analytische Mechanik der Schwingungen (Analytical Mechanics of Oscillations) Berlin, 1966.

[16] Thomson, W., and P. G. Tait: Treatise on Natural Philosophy, Volume I, Cambridge University Press, 1897.

[17] Zajac, E. E.: J. Aeronaut. Sci. 11 (1964), pp. 46–49.

[18] Chetayev, N. G.: The Stability of Motion (translated from the Russian original), Oxford / London, Pergamon Press, 1961.

[19] Ziegler, H.: Z. Angew. Math. Phys. 4 (1953) pp. 89-121.

[20] Grammel, R., and H. Ziegler: Ing. Arch. 24 (1956), pp. 351–372.

[21] Butenin, N.: Priborostroenije, (1963), No. 5, pp. 75–83.

[22] Contensou, P.: in "Kreiselprobleme" (Gyrodynamics) edited by H. Ziegler, Berlin / Göttingen / Heidelberg, Springer–Verlag, 1963.

[23] Magnus, K.: Advances in Aeronautical Sciences, Pergamon Press, Volume 1 (1959), pp. 507–523.

[24] Schuler, M.: Phys. Z. 24 (1923), pp. 344-350.

CONTENTS

Printed in the United States
By Bookmasters